Structuring the Journey to Work

Publications in the City Planning Series
Institute for Urban Studies
University of Pennsylvania
Robert B. Mitchell, Director

THE PLACE OF THE IDEAL COMMUNITY IN URBAN PLANNING
Thomas A. Reiner

HOUSING MARKETS AND PUBLIC POLICY
William G. Grigsby

EXPLORATIONS INTO URBAN STRUCTURE
M. M. Webber, J. W. Dyckman, D. L. Foley,
A. Z. Guttenberg, W. L. C. Wheaton, C. B. Wurster

STRUCTURING THE JOURNEY TO WORK
Howard S. Lapin

URBAN RENEWAL IN EUROPEAN COUNTRIES: ITS EMERGENCE AND POTENTIALS
Leo Grebler

Structuring the Journey to Work

by

Howard S. Lapin

Institute for Urban Studies
University of Pennsylvania

Philadelphia
University of Pennsylvania Press

Published in Great Britain, India, and Pakistan
by the Oxford University Press
London, Bombay, and Karachi

Library of Congress Catalogue Card Number: 63-15012

7424
Printed in the United States of America

Foreword

THERE IS NO NEED to belabor the importance of transportation to the life of a city, nor the extent of the problems created by an inefficient circulation system. One of the most significant measures of the adequacy of a transportation system is its ability to carry the enormous daily movement of persons to and from their work. Only as we come to understand much more fully the determinants of patterns of this work travel can we improve urban transportation. This volume is directed toward the development of analytical tools which will be of use in increasing that understanding.

This study was made possible by a grant to the Institute for Urban Studies from the Ford Foundation. The original manuscript was completed in 1959 and the methods of analysis that were developed as part of the study and which are presented here have already proven useful in transportation studies undertaken in various parts of the country. We are indebted to Mr. Lapin for this important contribution to our understanding of the structure of the urban community.

Robert B. Mitchell,
Director

In memory of Ben L. Goldberg,
who exemplified the highest

Acknowledgments

THE AUTHOR TAKES FULL RESPONSIBILITY FOR THE STATEments and conclusions made in this report; however, he wishes to express his gratitude to the following persons for their valuable aid in the present effort.

For suggestions, review, and constructive criticism he particularly wishes to thank Professors Robert B. Mitchell and William L. C. Wheaton of the University of Pennsylvania.

The U.S. Bureau of Public Roads and the Philadelphia City Planning Commission made possible special machine tabulations of work-trip characteristics for several cities, which yielded basic data for this study. For this aid and for helpful discussions the author wishes to thank these officers and staff engineers of the Bureau: S. T. Hitchcock, E. H. Holmes, John T. Lynch, and Gordon B. Sharpe. Similarly, he is indebted to Larry Reich, formerly of the Philadelphia City Planning Commission, for his substantial aid and support. Raymond Desjardins of the Metropolitan Toronto Planning Board generously furnished traffic data and suggestions of value to this study. George Britton and Paul D. Muffley of the Highway Planning Division, Pennsylvania Department of Highways, furnished unpublished and published highway traffic information of considerable help. The work of Hans Blumenfeld on transportation analysis and on urban form was found to be uniquely valuable.

Those who furnished publications, data, and other information of substantial aid included: Joseph D. Carroll, Jr.,

Nathan Cherniack, John Dyckman, Henry Fagin, Philip E. Geissal, Meredith B. Givens, Ernest Jurkat, Walter K. Johnson, Dwight D. Kelley, Louis Levine, David D. Longmaid, Mrs. M. E. Meyerson, Martin Meyerson, Herman E. Olson, Richard Overmyer, Wilfred Owen, Victor Pelz, Chester Rapkin, Frederick Sass, and F. Houston Wynn.

The author wishes to thank those who helped substantially in the preparation of this study. They include: Mrs. Grace Milgram, editor; Walter Kondo and Arthur Schwartz, who assisted him in preparing work-trip analyses from tabular data; Michael Rubenstein, who prepared the drawings; and Mrs. Mary C. Ellison and Mrs. Ruth Monheit, who typed the manuscript.

Contents

List of Tables

List of Figures

Introduction

COMMON TO VIRTUALLY ALL WAGE-EARNERS AND PROPRIETORS in American cities is the daily journey from home to work and back again. An easily observable characteristic of this travel is that most of it occurs at virtually the same time: a morning period of maximum volume of travel occurs in most cities from 7 to 9 o'clock, and in the evening another peak period of traffic flow in the opposite direction, away from the workplaces, occurs in the hours from 4 to 6 o'clock. The massive flows to and from workplaces contain many vehicles that alternately speed, stop, and crawl along, but more importantly, they include thousands or millions of people spending a considerable part of their days and of their lives in moving from one place of "residence" to another. These diurnal migrations are represented in the term, the "journey to work."

That most workers are employed away from their homes and travel between homes and workplaces at approximately the same times has considerable effect on capacity requirements of the circulation systems of cities. Further, the decisions made by workers and heads of families in selecting workplaces and in finding places to live, in the aggregate, have important effects upon the physical patterning of cities. Decisions on location also influence the availability of the labor force in various parts of the urban area. Not to overstate the case, however, it should be made clear that while the home-workplace relationship is clearly an operative factor at all times in providing limits, it may seldom be the prime factor in the choice of location.

From a statistical viewpoint, trips to work constitute the largest single grouping, by purpose, of all trips leaving urban residences. In metropolitan areas, trips from home to work and back again form generally about one third of all trips made, and up to half of those made to central business districts may be for a work purpose.

In a more important sense, work travel consumes much time and energy. A large proportion of such travel in metropolitan areas is arduous and long; many trips involve two or more changes of mode of travel. Two examples will help to give an idea of the scale of the journey to work and its costs. Estimates made for both the London central area and lower Manhattan indicate that time spent in the journey to work lengthens the workday by a gross amount of almost 20 per cent. Thus, for the approximately three million employed persons who travel into lower Manhattan each weekday, over half a million man-days are consumed in travel en route.

The journey to work is to be considered as an important factor of linkage among those areas in urban centers predominant in usage for residential, commercial, and industrial purposes. Wider understanding of the relation of homes to workplaces would aid in planning for the future organization of individual metropolitan areas. With evident trends to the decentralization of industry and the growth of low-density residential suburbs, decision for renewing the central areas of cities or for developing outlying subcenters of residence or employment cannot be made without assumptions concerning the work travel of individuals.

This study shows examples of the various means by which work travel has been or might be analyzed, so that cities may be improved as places for living and working, and the mobility of urban workers provide the greatest advantage at least cost. This study seeks to:

1. Indicate kinds of data required in the solution of problems concerning the journey to work.

2. Illustrate useful methods of analyzing the journey to work, primarily in terms of trip characteristics, factors affecting the generation of trips from residences, their confluence at workplaces, and patterns of interrelationship between residences and workplaces.

3. Improve predictability of characteristics of work-trip flows for traffic engineering and city planning purposes.

4. Point out some directions for further research to develop the power of the analytical tools described herein.

I

Previous Studies

STUDY OF THE JOURNEY TO WORK WAS OF MUCH GREATER interest to Europeans than to Americans in the first half of this century. Census and other materials showing extensive data on this subject date from at least as far back as 1896 in France and from 1900 in Germany and Switzerland. A major interpretive analysis of the journey to work, based upon railroad commutation statistics for Belgian workers, was published in 1910 by Ernest Mahaim. European studies on "Pendelwanderung" or "le mouvement alternant" or Japanese studies on places of work in relation to residences may be found in virtually every decade of this century. Most of these studies were based upon official census surveys in the respective countries. Great Britain has the distinction of having conducted two extensive census studies relating to the journey to work, comparable in detail, for the years 1921 and 1951, from which trends over time may be inferred. Other countries where official studies of the journey to work have been conducted include Holland, Denmark, Sweden, Venezuela, and Australia.

In the United States, the agency best able to compile information on the journey to work, the U.S. Bureau of the Census, as of 1959 had not undertaken official studies on it, excluding contract work performed for the U.S. Bureau of Public Roads in connection with surveys of the use of motor vehicles. However, formal studies concerning the journey to work do exist in this country. They include the real-property inventories of the early 1930's; studies of places of home and of work conducted by state departments of employment for several major cities and for at least one state; the war-worker surveys of World War II made in connection with gasoline rationing and programs for staggered hours; and the recent motor-vehicle-use surveys. Nonetheless, it is strange that this subject, which is of vital concern in planning for many phases of urban development, has been neglected in the official census studies. In 1957, this matter was called to the attention of the Bureau of the Census by a committee of the Population Association of America.[1] At the present writing (1959) it appears that two new questions in the Census of 1960 will relate to the journey to work.[2] A third new question will concern the availability of private automobiles.

The discussion below includes several of the recent major studies made in Great Britain because of their importance in calling attention to, as well as technically aiding, this aspect of urban research. Additionally, five basic approaches by authors in the United States to the study of urban travel are distilled and described.

[1] The Committee chairman was Gerald Breese, and Henry D. Sheldon of the Bureau of the Census served as liaison officer. Committee members included: Nathan Cherniack, William Grigsby, Howard S. Lapin, Harlan G. Loomer, Warren Lovejoy, Chester Rapkin, and Arthur Row. Organizations represented by these members were the Port of New York Authority, City of Philadelphia, Princeton University, and the University of Pennsylvania.

[2] Among the detailed questions to be submitted to a 25 per cent sample of the population will be these two:

a. "What city (or town) did he (the worker) work in last week?"

b. "How did he get to work?" (What principal mode was used?)

1.1. THE BRITISH STUDIES

It appears that several British reports have since 1941 redefined the problem and set the scale of study required concerning the journey to work. Among the important studies in England are the following:

1. *When We Build Again.* A study based on research into conditions of living and working in Birmingham. Published by the Bournville Village Trust. Research Director: C. B. Parkes. 1941.
2. Kate K. Liepmann, *The Journey to Work, Its Significance for Industrial Life,* London: Kegan Paul, Trench, Trubner and Co., 1944.
3. *London Travel Survey 1949.* London Transport Executive, 1950. (And similar subsequent studies.)
4. Ruth Glass, "London on the Move." *London Times,* June 18-19, 1956. And the more detailed report on the same study by John Westergaard, "Journeys to Work in the London Region." *The Town Planning Review,* April, 1957.

1.11. The Birmingham Report

The first report, on plans for the City of Birmingham, delved into the relationship between the locations of home and workplace. One chapter of this book is devoted to study of the journey to work. The part that the home-to-workplace relationship plays in organizational structure of the city was recognized and effort was made to utilize this knowledge in formulation of a postwar development plan for Birmingham.

1.12. Liepmann's Book

The second report, that by Liepmann, was a major force in calling attention to the city planning aspects of the journey to work. It is strongest in description of the social and economic functions and cost of work travel, but also presents detailed statistical data concerning work travel to a number of workplaces.

The data, from a variety of employers in England and on the Continent, do not provide a basis for any type of quantitative generalizations; however, the descriptive element should not be underrated. It is reasonable to believe that the excellent sampling surveys of the journey to work conducted in the London commuting area, beginning in 1948, are closely related to the pointed emphasis given work travel and its functional characteristics by Liepmann.

Liepmann considered that mobility of workers in selecting workplaces is an important factor in building labor force adaptability into a dynamic economic system. But she did not believe that adaptability should mean that the "greatest number of persons should actually undertake daily travelling" over considerable distances. Rather, she stated an important intangible quality concerning urban society, that "the value of mobility of labour, as of every form of freedom, lies in the potential as well as in the actual use made of it." Based on this premise, she postulated a goal of city planning: "what is needed is to bring alternative workplaces within daily reach of every earner," and proposed that specific consideration of the journey to work be an integral part of any plan for urban development "as a means of combining freedom with planning."

Liepmann generalized about urban form to the extent of supporting the idea of a regional pattern of distinct small towns about a central nucleus, rather than unchecked, formless sprawl emanating from the center city. She discussed particular density standards as a part of the concept of regional distribution of population; however, this part of the study might be described as tangential to the central description of work trips since the proposed density standards are not supported by detailed studies. In all, this work represents an important step forward in the consideration of the pattern of work trips as a part of the complex structure of urban areas.

1.13. London Travel Survey 1949

This was a workmanlike probability sampling study of the London travel market, with attention focused upon use of public transit and upon the journey to work. Numerical conclusions were presented along with materials from which sampling error could be estimated. This market study drew no inferences, presented no generalities. It indicated a concern about the journey to work by a public transit operator which has few, if any, counterparts on the western side of the Atlantic.

1.14. London on the Move

Like Liepmann, Ruth Glass, the director of a research group at University College, London, proceeded from study of the quantitative data to generalizations about planning policy. But while Liepmann was concerned mostly about density standards in the fringe or suburban areas, Glass directed attention to residential density standards in the central part of the metropolitan area, that is, the central area of London.

This study, described in a detailed journal article by Westergaard as well as in two newspaper articles by Mrs. Glass, was based upon comparative data for the London region from the 1921 and 1951 census reports. The information given located both the origin and destination by census area. Thus, it did not permit accurate estimates of trip lengths, but an approximation could be made with the implicit error of estimate varying with the size of the area units. With this limitation, it was possible to measure the changes over the 30-year interval in the accumulations of work trips in the London central area and in the places from which the trips were drawn. Both Liepmann and Glass were shocked by the trend toward increasing trip lengths and trip expense, and decried the then current planning policies, which promised to continue the undesirable trend.[3]

To what extent the critical writings of Mrs. Glass had an effect is not known; however, within a few months after her newspaper story, the London County Council announced a new planning policy to prevent further congestion in the central area.[4] The purpose of the new policy is to reduce the density of the daytime population and increase density of the resident population in Central London while, at the same time, preserving the city's character.

Here we have not only an example of a group of related studies driving toward recognition of a major problem, but also indications of positive official action taken in the direction recommended by the research. The sequence of events could be attributed to chance; but even so, the results are encouraging.

1.2. STUDIES IN THE UNITED STATES

Although well over a hundred major origin-destination (O-D) surveys have been conducted in this country since 1944, the number of interpretive studies of the journey to work has remained surprisingly low.

[3] Interesting enough, a third student of the journey to work in Great Britain noted the trend toward increasing lengths of travel and drew a conclusion different from that of either Liepmann or Glass. Jean Thompson concluded: "Clearly, planned decentralization is the only way out of this vicious circle, and if the social implications of the journey to work are considered, the drive for decentralization must be intensified." Her proposed solution was the development of greenbelt cities of "reasonable" size. "The Journey to Work: Some Social Implications," *Town and Country Planning*. XVIII (November, 1950), 441-446.

[4] ASPO *Newsletter*. November, 1957, p. 92.

It is omitting several reports of interest to do so, but on the present subject it is believed that the principal steps forward in this country may be reviewed by considering five distinct groups of studies: those by Douglas Carroll, Robert B. Mitchell and Chester Rapkin, Houston Wynn, the "gravity school," and the staff at Cornell University.

1.21. Carroll's Early Work

Carroll's doctoral thesis at Harvard University, which preceded his work in Detroit and Chicago, is important in a study of the evolution of viewpoints on the journey to work.[5] Carroll examined a large number of transportation surveys for individual cities, as well as a major statewide survey. The Massachusetts data on commutation and home-work relationships of industrial employees were compiled by the State Planning Board as a wartime measure to study means for conserving tires and gasoline without hindering production. The figures, concerning trips of 200,000 workers, were from 268 plants in small communities and the Boston Metropolitan Area, representing 25 per cent of the industrial workers throughout the state. Carroll studied differences in travel distance and mode with size of plant employment and with size of town or city where the plant was located, and made inferences concerning the apparent willingness of workers to move closer to workplaces. Where possible he sought travel patterns according to type of industry.

From these efforts, based on wartime and pre-World War II data, Carroll proposed a theory of the "spatial relationships of homes to workplaces" and sought to draw

[5] J. Douglas Carroll, Jr., *Home-work Relationships of Industrial Workers* (Unpublished thesis, Harvard University, 1950). Carroll's later work as director of the Detroit and the Chicago Transportation Studies has broken new ground in urban travel analysis, but the materials described here are considered sufficient to indicate the distinctive approaches to study of the journey to work.

their implications for city planning. The hypotheses suggested are: (1) "Forces are in operation tending to minimize distances between home and place of work," and (2) "The concentrative effect of these forces is an important factor conditioning total residential arrangement of urban populations."

Three general statements were made to describe spatial patterns of homes and workplaces:

1. "Total urban area population is residentially distributed about the central business district of the principal city."

2. "Residential distribution of persons employed in central districts tends to approximate that of the entire urban area population."

3. "Residences of persons employed in off-center workplaces are concentrated most heavily in the immediate vicinity of the place of work." [6]

Carroll's initial study on the subject is commendable—even though some of the inferences made depend solely upon surveys of the 1920's and 1930's, when the survey data were considerably less comparable from city to city than they are at present.

A more recent study of home-to-work trips in the vicinity of Boston by A. J. Bone and Martin Wohl might be considered as somewhat of a sequel to Carroll's work with Massachusetts data.[7] A postcard survey was conducted among employees at workplaces located near a new major highway facility, Route 128, which is located on the periphery of the Boston metropolitan area. The intent of Bone and Wohl was to study the effects of a new highway. Com-

[6] "The Relations of Homes to Work Places and the Spatial Pattern of Cities," *Social Forces*, XXX (March, 1952), 271-282.

[7] A. J. Bone and Martin Wohl, "Economic Impact of Route 128," *Traffic Engineering*, XXVIII (July, 1958), 11-17, 36-39. An article adapted from a paper presented at the Highway Research Board, Washington, D.C., on January 8, 1958.

parison of this work and Carroll's data might be of value in assessing trends in work trips over the years between the surveys of 1942 and 1958.

One impression is gained from the study, that perhaps Carroll assumed the location of home in relation to workplace depended too heavily on the characteristics of the journey to work.[8] Thus time-distance, convenience, and cost are the characteristics of the journey to work that affect residence in relation to work, but within the limits of a given length and cost of trip are there not a great many variables, such as the character of the neighborhood, school accessibility, playgrounds, religious organizations, and rents and taxes, which are major forces in location of the residence?

Another point in regard to the concept of the minimization of distance is desired: A *limiting* minimum distance might also be noted in most of the area distributions of residences about workplaces. A question can be asked as to the existence of a repulsion effect operating at a distance of up to one half or three fourths of a mile. (Also to be considered, of course, is the dearth of residences close to plants in major industrial land-use zones.) It is fair to say that Carroll's initial efforts on this subject helped to advance in this country the concept of the journey to work as an important factor in the structuring of cities.

1.22. Mitchell and Rapkin

The work of these authors represents a major contribution to the theory of urban travel in relation to city devel-

[8] Influenced by the Zipf concept of minimization of energy in travel, Carroll sought to show that "any change in workers' homes (locations) will be in the direction of reducing work travel distances" (Thesis, p. 162). George K. Zipf presented his concept of the social effects of minimization in two works quoted by Carroll: *Human Behavior and the Principle of Least Effort* (Cambridge, Mass.: Addison-Wesley Press, 1949); and "The Hypothesis of the Minimum Equation as a Synthesis," *American Sociological Review,* XII (December, 1947).

opment.[9] It broke down the analysis of urban travel into sufficiently refined components that traffic could be related to the specific land-use areas from which it originates or at which it terminates. In addition, a terminology was developed to facilitate handling of the more finely considered elements in urban transportation problems. The concepts of systems and structures of movement (as discussed in Appendix C, Section 20) illustrate only two of a number of proposed analytic approaches.

The concept of *systems* of urban traffic led to formulation of a large part of the presentation of the present report. The proposal to divide further the work trip of the O-D surveys (Appendix C, Section 20) and the attempt to do so by special runs and tabulations of the work-trip deck were both by-products of the consideration of urban traffic in terms of its component systems.

The present report has not adopted the concept of trips proposed by Mitchell and Rapkin, since it is dependent upon existing O-D trip analysis and the new definition was not found adaptable for use with available data. The omission here does not argue against its use in a totally new survey. The meaning of the term trip as considered in the present report is discussed in Appendix C. The proposed new definition would have incorporated the movement of an individual between bases (or loci). The bases are places where individuals are functioning members, as at work, at home, or school, etc. Thus, the trip consists of several legs by this new definition, and the legs of the trip may be by various modes and terminate between bases at various intermediate temporary stops along the way. This definition has the advantages of bringing into consideration the walking part of each movement between bases, of eliminating the confusion about "change-travel-mode,"

[9] *Urban Traffic, A Function of Land Use* (New York: Columbia University Press, 1954).

and of showing more clearly the relationship between lesser and major purposes of each movement. In using the existing definition, subsidiary trip purposes either appear as major purposes or are lost altogether.

As well as providing a framework for studies of trips generated by various land uses, the Mitchell-Rapkin book sought the relationships among trip generators. Thus a concept of linkage among establishments was proposed, which looked at land-use interrelationships in terms of urban travel.

The present study is considered to be an example of the kind of research on the systems of urban traffic called for in Chapter IX of *Urban Traffic, A Function of Land Use.*

1.23. Houston Wynn

Wynn has contributed to the present understanding of the patterns of urban work trips in studies conducted for a number of cities. He has given considerable attention to the separation of work trips from all others, and to use of auto and public transit for the two major groups of trips, those for work and nonwork purposes. He has sought per capita trip-generating factors to apply to areas of varying land use, so that a synthetic pattern of travel might be created to forecast future travel better.

1.24. The Gravity School

In recent years, use has been made in urban traffic studies of the analogy between Newton's law of universal gravitation and empirical findings of the interactance among groups of people over varying distances. Interactance increases directly as a function of the product of the population size of the groups, but decreases with increasing distance between the centers of population. (Interaction between area i and area j may be expressed as

$I_{ij} = kP_iP_jf(d_{ij})$, where P represents population of the respective areas, d the distance between their centroids, and k is a qualifying multiplier.)

A short but wide-ranging review of this concept and its historical evolution from the first half of the nineteenth century was prepared by Carrothers.[10] The usage by Carrothers, and most other authors he mentions, pertains to trips for all purposes in or between cities. A smaller number of writers have attempted to relate the gravity concept to analysis of work trips. The clearest and most useful formulations of this type have been included in materials published by the Detroit Metropolitan Area Traffic Study (DMATS).[11] Two others have worked with adaptations of the DMATS materials. These were Alan Voorhees and myself.[12] The second of these two references is briefly described in footnote 32 of Chapter IV. The method it proposes was utilized by Robert Coughlin of the Philadelphia City Planning Commission in preparing a first-approximation projection of work trips and of total vehicle trips.[13]

Another important concept from the students of social physics is that of the potential of all other points in a field of interactance, which act upon a particular point within the field. Thus, the potential at an area *i* is represented by the sum throughout the field of influence of the quantity being measured (say population) at each area other than

[10] "An Historical Review of the Gravity and Potential Concepts of Human Interaction," *Journal of the American Institute of Planners*, XXII (Spring, 1956), 94-102.

[11] In particular, see Volume I, *Report on the Detroit Metropolitan Area Traffic Study*, 1955. J. Douglas Carroll, Jr., Director.

[12] Alan Voorhees, "Forecasting Peak Hours of Travel." A paper presented at the Highway Research Board, Washington, D.C., January 8, 1958.

Howard S. Lapin, "The Analysis of Work Trip Data," *Traffic Quarterly*, XI (April, 1957), 278-292.

[13] Robert Coughlin, *A Projection of Vehicle Trips within Philadelphia* (Philadelphia City Planning Commission, 1957 [Mimeo]).

i, with each quantity multiplied by a relevant function of the respective distance of its centroid from *i*.

It may be expected that the further development of these two fundamental concepts, in analyses of trip generation and trip distribution, will lead to major advances in understanding of patterns of urban travel. Discussion relevant to this subject is to be found in Chapter IV.

1.25. The Cornell Studies

Several important contributions to the study of the journey to work have been prepared by faculty members of Cornell University.[14] Whereas most other available studies concern travel within or to great urban agglomerations, the Cornell publications are concerned with the journey to work in and around the relatively small towns of upstate New York and from the rural areas.

Special notice in the Adams-Mackesey report was taken of "long-distance commuters," those workers who travel more than 20 miles each day to their place of work. Such long work trips were almost all (98 per cent) made by automobile.[15]

The Adams-Mackesey study makes the point that there no longer exists a "clear cut occupational distinction between so-called urban or suburban residents and residents

[14] Leonard P. Adams and Thomas W. Mackesey, *Commuting Patterns of Industrial Workers* (Ithaca: Cornell University Housing Research Center, 1955).

Glenn H. Beyer, *Housing and Journey to Work* (The Patterns of Rural Families in Monroe County). Cornell University Agricultural Experiment Station, August, 1951.

[15] Other sources of information on commuting to manufacturing plants from rural open country are these:

Howard E. Conklin, "The Rural-Urban Economy of the Elmira-Corning Region," *The Journal of Land and Public Utility Economics*, XX (February, 1944).

George P. Stevens, Jr.: "Sample Study of Residential Distribution of Industrial Workers in an Urban Community," *Land Economics*, XXVIII (August, 1952), 278-283.

of the open country or small villages."[16] They found that in some parts of the northeast region from two thirds to three fourths or more of the open-country dwellers gained their principal income at nonfarm work. They also found that only "10 to 15 per cent of those who work off the farms try to operate farms on a part-time basis."[17]

The authors believe that, at least in New York State, future employment opportunities will be relatively more numerous in plants on the peripheries of established major industrial centers, and that workers will show a continuing tendency to establish their homes in the suburban area, in close-by villages, or in the open country. They conclude that, in general, the numbers of long-distance commuters will not decline, but do not further infer a pattern for this type of work travel.

Research on the journey to work conducted at the Housing Research Center of Cornell University was based on three sources of data: those collected during World War II concerning travel of workers to defense plants,[18] labor market studies in years subsequent to World War II, and figures concerning commuting in several upstate New York communities in 1951-1952. Adams and Mackesey concluded from the several bodies of data that commuting distance is dynamic over time and subject to considerable variation among communities of a single geographic region. However, they noted a broad pattern to the extent that data on trips of hundreds of thousands of wartime workers indicated that three fourths of the workers resided within ten miles of their workplaces and about 95 per cent within 20 miles.

[16] *Op. cit.*, p. 71.

[17] *Ibid.*

[18] The principal sources of data on wartime transportation were: Theodore M. Matson, *War Worker Transportation* (New York: Institute of Traffic Engineers, 1943).

U.S. Public Roads Administration, *War Worker Transportation*, March 31, 1943 (Mimeo).

Beyer interviewed families in 601 dwellings in the non-urban sections of Monroe County, New York. He found distinct income, industry-of-occupation, age, and housing patterns among the commuters from open-country areas. The commuter families were about equal in number to the others in the rural area studied. The majority (61 per cent) of the commuting workers were engaged in manufacturing. Median income of heads of households in commuter families was $3,500 as compared with $2,300 for heads of households in the other families. Of the commuters, 94 per cent got to and from work as drivers or as passengers in automobiles. The median trip length was 12 miles. The one-way trip length exceeded ten miles for nearly two thirds of the commuters.

What emerges from the rural studies is a picture unfamiliar to urban planners and traffic engineers. From the point of view of studying the large traffic flows in cities, the relatively small percentage of trips from external areas does not seem to call for much attention. Yet, considered over the rural countryside only, we find the large employment centers, particularly the factories, to be very important in the present-day rural economy. For over half the cases tested, the families in the open-country area must be considered residents of an exurbia, the hinterland of metropolitan dependence.

II

The Character of Work Travel

WORK TRAVEL TAKES PLACE IN AN ENVIRONMENT WHERE TRIPS are made for a great variety of purposes. It is the trips for all purposes that have until quite recently been studied en masse in terms of desire-lines in urban traffic surveys. The present study deals only with the systems of work travel, but it is nonetheless important to observe the relationships between these and total urban travel. The concept of relationship is vital for qualitative and quantitative reasons: first, to give an idea of scale to the discussion of work travel, and secondly, to provide data from which multipliers might be derived to expand to a more complete basis information which may be derived solely from systems of work travel. The latter purpose applies particularly in those cities where only work travel has been found measurable for budgetary reasons (as in Los Angeles and Toronto), but elsewhere as well because of the greater completeness of its measurement as compared with trips for other purposes.

Generalizations about work trips and their patterns must be based upon assumptions concerning the statistical behavior of large numbers of individuals. Formulation of mathematical statements concerning distributions of origins in residential areas, in what appears to be a kind of nonfree-will behavior, has a flavor of abhorrence to most people. But paradoxically, the probabilistic bases of generalization about such behavior assume that each individual has essential freedom of choice. The financial health of the large insurance companies is a demonstration that such generalization works. That behavior is describable in this way need not be redemonstrated here.

The mathematical generalizations on travel among a number of zones may follow the so-called gravity format, or a variation thereof, or proceed from an independent empirical base. An abundant literature exists on the subject of patterns of human interaction over geographic space. A repetition of these sources will not be given here. Rather, a few references, cited in the footnote below, represent a major coverage of the more successful approaches to generalization about intraurban trips for all purposes.[1] The Carrothers article will be found particularly interesting for its general summary of the gravity formulation in analysis of urban travel and for its extensive bibliography.

This chapter concerns measurable aspects of the journey to work as an element of urban traffic: its proportions relative to other systems of movement, and such numerical characteristics as time- and mile-distance, modes selected, variations in length of trip with the population class of the city, and measures of ride-sharing in automobile work trips. In addition, the chapter introduces several of the variables associated with differences in structures of urban work travel.

An important qualification should be observed in the

[1] Gerald A. P. Carrothers, *op. cit.;* Highway Research Board, *Travel Characteristics in Urban Areas.* Bulletin 203. December, 1958. Contains seven papers on urban travel. Bulletin 224 of the HRB is also relevant; Sam Osofsky, "The Multiple Regression Method of Forecasting Traffic Volumes," *Traffic Quarterly,* XIII (July, 1959), 423-445; Alan M. Voorhees, ed. "Land Use and Traffic Models: A Progress Report." Special issue of the *Journal of the American Institute of Planners,* XXV (May, 1959). Contains nine useful articles.

presentation of these variables. For the descriptive purposes of this chapter, several measurable characteristics of work trips are shown in their relationship to such associated factors as the income level of travelers, or the place of travelers' residence within the metropolitan area. Because of the type of data available, only a few of the factors are presented in each of the examples cited. Examples of differences in work-trip characteristics with variations in relevant factors help to illuminate the facets of a many-sided problem. The effects, that is, the details of urban travel behavior useful to transportation planning, together with the associated factors, are to be studied by the various techniques of association analysis. When applicable factors have been isolated and their relations to effects described, then appropriate microarea data of these factors may be employed in predicting patterns in trip generation and in the distance traveled. These matters are discussed in the succeeding two chapters and in Chapter VI. It is believed that the examples cited here will be helpful in specific problems in finding useful relationships among important variables.

2.1. THE PROPORTION OF WORK TRIPS TO ALL TRIPS

Trips for work purpose range from one sixth to one third of all vehicular trips recorded in internal area surveys, with an average volume of about one fourth of the total.[2] The largest group of trips by objective are those made to home, which for 15 cities of from 75,000 to 1,500,000 population ranged from 37 per cent to 48 per cent and averaged 43 per cent. These include those made from work to home, as well as trips made from the great variety of other possible activity areas. There was no correspondence

[2] See Table 6-6, Matson, Smith, and Hurd, *Traffic Engineering* (New York: McGraw-Hill, 1955).

in these sets of figures, for either work or home trips, between the proportions of trips by purpose and population size of city.

The proportion of work travel, in the figures above, is described in terms of the total of two systems of trips from internal-area origins: the home-work trips, and the other-work trips. In comparing the volume of work trips from city to city, the two systems should not be separated since together they represent the totality of internal-area work-purpose trips. Thus, in the internal area, the journey to work, that is, home-work plus work-home trips, represents something less than double the proportion of work-purpose trips, depending upon the volume of other work trips. (See Appendix C for an explanation of the terms describing systems of trips.)

One reason for the wide variations in the reported percentage of work trips as part of all internal trips is the result of differences in definition and tabulation among the surveys. Thus, trips to home are sometimes included in the tabulated total and sometimes omitted. Other smaller groups of trips that similarly may be included or omitted are the change travel mode and serve passenger trips. In addition, and confusing the picture by another degree, work trips may also be reported in terms of their proportion of trips to ultimate, nonhome destinations.

Thus, in data for Philadelphia internal trips, we have three possible proportions, as shown in Table II-1. The data are for the year 1947, so the absolute figures are no longer valid, but for our purpose the proportional relationships hold their meanings over time. It is seen that work trips represented 25 per cent of all trips, 39 per cent of all trips other than those to home, and fully 56 per cent of trips made for an ultimate purpose other than getting home.

Let us consider another scale factor, the proportion of

TABLE II-1

DISTRIBUTION OF INTERNAL-AREA TRIPS BY PURPOSE
PHILADELPHIA-CAMDEN AREA, 1947

Trip purpose	*All modes (1,000's of trips)*	*Work-trip Percentage*
All trips	3,548	25
To home	1,305	—
All trips, less to home	2,243	39
To change travel mode	628	—
To serve passenger	47	—
To ultimate purpose (except to home)	1,568	56
Work	871	—
Business	83	—
Medical, dental, and school	108	—
Social, recreational, and eat meal	323	—
Shopping	183	—

work trips to total trips in the vicinity of a workplace destination. This proportion is most readily measurable within the confines of designated central business districts (CBDs). As measured by O-D studies in large cities, the work trips appear to represent from 50 to about 55 per cent of trips made to destinations in the designated CBDs. Reasons for the variation in this proportion lie in the differences from city to city in scale and influence of the CBD relative to other local centers of commercial activity, and in the necessarily arbitrary assumptions made in the delineation of the CBD areas.

It should be noticed that the trips discussed are those made to destinations in the CBD. A great many more trips may be made to and *through* the CBD, so in order to make city-to-city comparison of this type of proportion more meaningful, only those trips have been considered which have goal activities within the CBD. An idea of the relative number of trips to a CBD and trips into and through the district can be gained from a Philadelphia study, where it was found that the vehicular trips for all purposes into and destined to the central business district comprised half of the total vehicular trips made into the district.

The fact that the proportions of work travel relative to the total number of trips and destinations within the CBD may be different leads inferentially to another concept, that of the highly localized variations in proportion of work trips to total trips which are to be found within the areas of influence of individual workplaces located outside the CBD. This local "field phenomenon" does occur about workplaces wherever they may be located. It is most discernible for the CBD, which in most urban areas represents the largest accumulation of work trips, but it may be found about other workplace destinations, whose ratios of work travel to total travel vary in a local pattern in accordance with their respective volumes of employment.

Most of the proportions given for work trips relative to total trips are for the internal area only. The explicitness in qualifying data as being applicable in either internal or external surveys may leave the question of the extent to which these proportions are changed by inclusion of both the internal and external trips. Two factors condition the answer to the question:

a. The external trips do not bear a constant relationship to the total of internal and external; but in urban areas of over one hundred thousand population, the proportion may be assumed to be well under one fourth of the total.

b. Trips to and from work are customarily counted as work trips in the external survey, as described in Appendix C. In the internal survey, only trips to work are so counted. (This distinction derives from the less precise treatment usually accorded external trips.)

From the two factors above, it may be assumed that work trips in the total of internal plus external trips will be of a somewhat higher proportion than for the internal survey trips only. Little in representativeness of data is lost in intercity comparisons of the proportions of internal work trips only, while comparability is enhanced. Table II-2 shows proportional distributions of external trips as measured in two major metropolitan areas.

Worth noting is the comparability of the proportional distributions in the two metropolitan areas for the last three of the five listed trip purposes (Table II-2). The similarity of the sums of proportion of trips for work and business (50.2 and 56.5 per cent) raises the question of whether or not the same meanings for the two trip purposes were employed in the interview processes. (It does not seem reasonable that the distinction between auto occupants and auto drivers could account for the proportional differences noted in work and business purposes of trips.) Perhaps the percentage differential indicates a kind of classification problem worthy of further study.

2.2. LENGTH OF TRIP

Work-trip lengths can be measured both in units of duration of time and of miles of travel. The consensus among transportation planners is that variations in time duration have more significant effects upon travel behavior in cities than do variations in miles. Even so, the mile-distance data are more generally available for traffic analysis purposes, being presented in the motor-vehicle-use studies and derivable from the many postwar urban O-D surveys. However, comprehensive time-distance studies have been prepared in connection with several major housing market investigations, as well as in some of the more recent and more intensive traffic surveys. Because of the

TABLE II-2

PURPOSE OF EXTERNAL TRIPS BY AUTO

	Auto occupants Minneapolis-St. Paul[a]			Auto drivers Philadelphia-Camden[b]		
Purpose of trip	*Number*	*Per cent*	*Per cent (sub-total)*	*Number*	*Per cent*	*Per cent (sub-total)*
Work	36,151	27.0	50.2	150,433	46.6	56.5
Business	31,036	23.2		32,075	9.9	
Social, recreational	48,626	36.4		95,186	29.5	
Shop	8,012	6.0		19,151	5.9	
Other	9,910	7.4		25,799	8.1	
TOTAL	133,735	100.0		322,644	100.0	

Sources:

[a] Minnesota Department of Highways, *St. Paul-Minneapolis Traffic Survey*, 1949.

[b] Pennsylvania Department of Highways, *Philadelphia Area Traffic Survey*, 1955. (Note: Work trips represented 43.6 per cent of external auto-driver trips in 1947.)

greater availability of mile-distance data, such compilations appear more frequently in the present report, but both types of data are interpreted, and methods are explored for their further utilization.

2.21. *Work-Trip Length in Miles Related to City Population*

From available information it is possible to formulate a rough picture of the variation in one-way length of work trips with differences in population size of cities. Two sampling surveys are available for different periods in time (Table II-3). One was conducted in 1941, consisting of 2,491 interviews based on carefully selected probability samples from cities throughout the United States. The second was conducted in the summer of 1951 in six states as a part of the motor-vehicle-use study program of the U.S. Bureau of Public Roads. Both surveys counted trips between homes and workplaces by all modes, including walking. To aid in comparisons between the surveys, data are given for the median length and 75th percentile trips as indicated from the cumulative distributions. Available data do not permit meaningful consideration of the 25th percentile level for the several city size classes for 1951.

TABLE II-3

LENGTH OF WORK TRIP BY POPULATION OF CITY

Population (in thousands)	Length of work trips — 1941 Median (in miles)	1941 75th percentile	1951 Median	1951 75th percentile
5 to 25	0.8	1.9	1.2	2.3
25 to 100	1.6	3.4	1.9	3.1
100 and more	3.2	7.6	3.3	5.2
100 to 500	2.0	4.5	—	—
500 and more	4.7	9.0	—	—

Source: See Figures II-1 and II-2.

It is believed that some time trends may be inferred from the two sets of data for the cities with populations of 5,000 to 25,000 and 25,000 to 100,000. It does not seem likely that the cities of over 100,000 population in the six states studied could be considered statistically representative of all U.S. cities in this size group, because of the scarcity of very large cities in the area sampled. In the 1941 survey, data were given for the population class of over 500,000. No comparable category is available for 1951. With this caution, the reader may infer what he will from

TABLE II-4

PERCENTAGE DISTRIBUTION OF TRIPS BETWEEN HOMES AND WORKPLACES; BY MILEAGE DISTANCE AND BY POPULATION OF CITY

SELECTED CITIES IN THE UNITED STATES, 1941

	Distance									
Population (in thousands)	*Up to ½*	*½ to 1*	*1 to 2*	*2 to 3*	*3 to 5*	*5 to 10*	*10 to 20*	*Over 20*	*Does not work any one place*[a]	*Number of inter-views*
Over 500	14	4	5	5	12	23	11	3	23	742
100 to 500	16	9	16	12	11	12	5	2	17	570
25 to 100	16	14	16	11	7	5	5	4	22	601
5 to 25	28	16	11	3	3	5	2	4	28	578
WEIGHTED AVERAGE	18	12	12	8	8	11	5	3	23	2,491

Source: National Survey Research Investigation, reported in *Urban Planning and Public Opinion*, by Melville C. Branch, Jr. (The Bureau of Urban Research, Princeton University, September, 1942).

[a] Data of indeterminate destination were eliminated in the calculations of the cumulative percentage distributions which were prepared for plotting these findings graphically.

TABLE II-5

Distribution of Gainfully Employed Workers Traveling from Home to Work by Mode of Travel, Population of City of Residence, and Distance Six States,[a] Summer, 1951

Population and one-way distance (in miles)	All workers		Percentage distribution of workers by mode of travel					
	Number	Per cent	All modes of travel	Passenger car	Public transportation	Passenger car and public transportation	Walk	All other means and not reported
5,000-25,000								
0.1- 0.9	188,558	44.7	100.0	47.2	1.7	0.5	48.4	2.2
1.0- 1.9	115,541	27.4	100.0	80.3	6.1	1.8	10.3	1.5
2.0- 2.9	31,351	7.4	100.0	80.5	10.8	1.7	2.0	5.0
3.0- 4.9	19,661	4.7	100.0	85.7	10.6	—	3.7	—
5.0- 9.9	21,754	5.1	100.0	90.0	4.2	2.4	—	3.4
10.0-19.9	30,776	7.3	100.0	95.2	4.0	—	—	0.8
20 and over	14,593	3.4	100.0	90.0	8.4	—	—	1.6
TOTAL REPORTED	422,234	100.0						
NOT REPORTED	53,917		100.0	38.5	1.5	0.4	19.7	39.9
ALL DISTANCES	476,151		100.0	64.4	4.2	0.9	24.2	6.3
25,000-100,000								
0.1- 0.9	106,180	25.0	100.0	38.8	5.2	1.0	53.0	2.0
1.0- 1.9	119,120	28.1	100.0	68.2	17.9	0.3	12.0	1.6
2.0- 2.9	89,032	21.0	100.0	71.0	26.2	0.5	1.5	0.8
3.0- 4.9	54,456	12.8	100.0	73.5	23.1	—	0.8	2.6
5.0- 9.9	34,868	8.2	100.0	87.3	8.6	1.0	1.3	1.8
10.0-19.9	12,517	3.0	100.0	87.2	6.7	3.4	—	2.7
20 and over	8,133	1.9	100.0	71.2	9.1	15.6	—	4.1
TOTAL REPORTED	424,306	100.0						
NOT REPORTED	22,248		100.0	34.5	13.1	—	17.9	34.5
ALL DISTANCES	446,554		100.0	62.8	15.7	0.9	17.2	3.4
100,000+								
0.1- 0.9	95,792	14.5	100.0	26.0	14.2	1.1	58.0	0.7
1.0- 1.9	98,828	14.9	100.0	40.1	49.1	0.8	8.7	1.3
2.0- 2.9	105,538	16.0	100.0	43.5	51.8	1.7	1.7	1.3
3.0- 4.9	190,032	28.8	100.0	48.7	45.7	3.6	—	2.0
5.0- 9.9	122,466	18.5	100.0	62.5	35.2	1.2	0.1	1.0
10.0-19.9	38,739	5.9	100.0	73.6	25.9	—	—	0.5
20 and over	9,057	1.4	100.0	65.7	10.6	19.9	—	3.8
TOTAL REPORTED	660,452	100.0						
NOT REPORTED	55,455		100.0	32.7	31.2	3.0	7.8	25.3
ALL DISTANCES	715,907		100.0	46.4	38.4	2.2	9.9	3.1

Source: T. A. Bostick, R. T. Messer, and C. A. Steele, "Motor-Vehicle-Use Studies in Six States," *Public Roads*, XXVIII (December, 1954), 111.

[a] Arkansas, Louisiana, North Dakota, Oklahoma, South Dakota, and Wisconsin.

the distributions for the two time periods for 100,000 plus cities, as indicated in Tables II-4, II-5, II-6, and II-7 and shown in Figures I and II. The figures show cumulative distributions on probability coordinates.[3]

TABLE II-6

DISTRIBUTION OF GAINFULLY EMPLOYED WORKERS TRAVELING FROM HOME TO WORK IN PUBLIC TRANSPORTATION, BY POPULATION OF CITY AND DISTANCE

SIX STATES,[a] SUMMMER, 1951

Population and one-way distance (in miles)	Total number of workers	Using public transportation: Number	Using public transportation: Per cent of all workers	Per cent distribution	Vehicle miles	Transit trip length: Average (in miles)	Transit trip length: Median (in miles)
5,000-25,000							
0.1- 0.9	188,558	3,205	1.7	16.78	1,603		
1.0- 1.9	115,541	7,048	6.1	36.91	10,572		
2.0- 2.9	31,351	3,386	10.8	17.74	8,465		
3.0- 4.9	19,661	2,084	10.6	10.91	8,336		
5.0- 9.9	21,754	914	4.2	4.79	6,855		
10.0-19.9	30,776	1,231	4.0	6.45	18,465		
20+	14,593	1,226	8.4	6.42	30,650		
TOTAL	422,234	19,094	4.5	100.00	84,946	4.45	1.90
25,000-100,000							
0.1- 0.9	106,180	5,521	5.2	8.20	2,761		
1.0- 1.9	119,120	21,322	17.9	31.67	31,983		
2.0- 2.9	89,032	23,326	26.2	34.64	58,315		
3.0- 4.9	54,456	12,579	23.1	18.68	50,316		
5.0- 9.9	34,868	2,999	8.6	4.46	22,493		
10.0-19.9	12,517	839	6.7	1.25	12,585		
20+	8,133	740	9.1	1.10	18,500		
TOTAL	424,305	67,326	15.9	100.00	196,953	2.93	2.29
100,000+							
0.1- 0.9	95,792	13,602	14.2	5.28	6,801		
1.0- 1.9	98,828	48,525	49.1	18.83	72,788		
2.0- 2.9	105,538	54,669	51.8	21.20	136,673		
3.0- 4.9	190,032	86,845	45.7	33.70	347,380		
5.0- 9.9	122,466	43,108	35.2	16.73	323,310		
10.0-19.9	38,739	10,033	25.9	3.89	150,495		
20+	9,057	960	10.6	0.37	24,000		
TOTAL	660,452	257,742	39.0	100.00	1,061,447	4.12	3.28

Source: T. A. Bostick, R. T. Messer, and C. A. Steele, *op. cit.*

[a] Arkansas, Louisiana, North Dakota, Oklahoma, South Dakota, and Wisconsin

[3] The ordinate or percentage data are plotted on a probability scale, a scale along which a grouped array of data appears as a straight line if the distribution may be represented by the normal curve. Abscissa data may be plotted according to a Cartesian (uniformly spaced) scale, or for some exponential distance functions, according to a logarithmic scale.

TABLE II-7

LENGTH OF WORK TRIP AS A FUNCTION OF MODE AND CITY SIZE

SIX STATES,[a] SUMMER, 1951

Mode of Travel	*Population size of city* 5,000 to 25,000	25,000 to 100,000	100,000 plus
Public transportation			
Per cent of reporting workers	4.5	15.9	39.0
Length of trip, average (miles)	4.45	2.93	4.12
median (miles)	1.90	2.29	3.28
Passenger automobiles			
Per cent of reporting workers	67.7	64.3	47.5
Length of trip, average (miles)	4.30	3.66	5.44
median (miles)	1.58	2.22	4.00
All modes[b]			
Per cent of reporting workers[c]	100.0	100.0	100.0
Length of trip, average (miles)	3.35	3.12	4.46
median (miles)	1.20	1.89	3.31

Source: T. A. Bostick, R. T. Messer, and C. A. Steele, *op. cit.*

[a] Arkansas, Louisiana, North Dakota, Oklahoma, South Dakota, Wisconsin.

[b] Includes the following mode categories: combination of auto and public transportation, walking, and "all other means and not reported."

[c] "Reporting workers" refers to those who reported length of trip. The total of "reporting workers" includes some who failed to report on modes used.

In the smallest class of cities considered (5,000 to 25,000 population) the median trip length may be seen to have increased from 0.8 to 1.2 miles from 1941 to 1951. Similarly, at the 75th percentile level, the trip length increased from 1.9 to 2.3 miles over the intervening decade. The tendency toward increased work-trip length in this group of cities seems to have been experienced by most of the residents over the period 1941-1951. There was an exception, however. Six per cent of the 1941 trips, those of the greatest distance, were longer than the top six per cent of 1951. Whether or not these data may be read so finely is another question; selection of an exact percentile of intersection of the curves is probably not justifiable. Let it suffice to say that all but the very longest trips increased in length in this class of city over the intervening decade.

Median trip lengths also appear to have increased in cities of intermediate size (25,000 to 100,000) from 1941 to 1951, although the absolute increase (0.3 miles) is less than the absolute increase (0.4 miles) for the median trips in smaller cities. In spite of the increase in median distance the upper 30 per cent in 1951 were found to be shorter than the 30 per cent of trips in 1941 of greatest length.

For those interested in the progression of change with size of city, and mindful of the caution given above, it is pointed out that the increase in the median trip length in the same period for the cities of over a hundred thousand population was practically negligible, being measured at 0.1 mile (3.2 to 3.3 miles), and 46 per cent of trips in 1941 of greatest length were found to exceed in distance the upper 46 per cent of 1951 trips. The finding of a relatively constant median trip length for cities of this class may be compared with results of an analysis of time-lengths of trips to work by Philadelphia residents in the years 1934 and 1956. Here the median was found to have remained constant over the 22-year period, while the proportions of trips of shorter and longer distances increased over the period. (Sources used by the author are those given in note *a*, Table II-11, and in Table II-8.)

Figure I

CUMULATIVE PERCENTAGE DISTRIBUTION OF LENGTHS OF WORK TRIPS, BY SIZE OF CITY - 1941

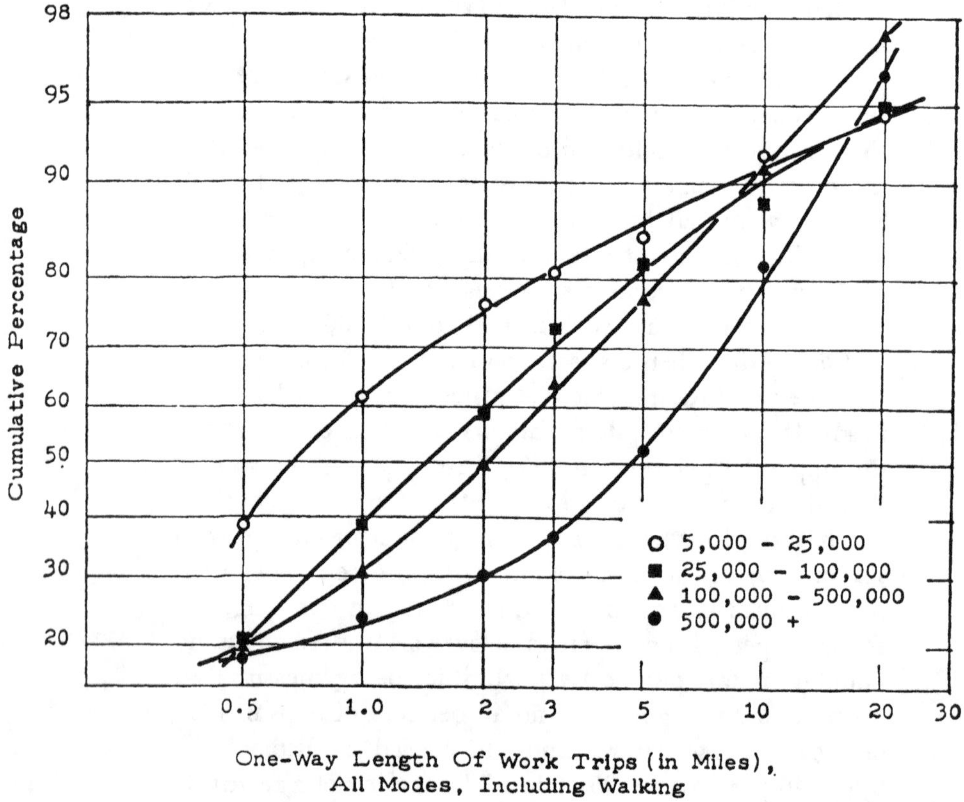

Figure II

CUMULATIVE PERCENTAGE DISTRIBUTION OF LENGTHS OF WORK TRIPS, BY SIZE OF CITY, IN SIX STATES — 1951

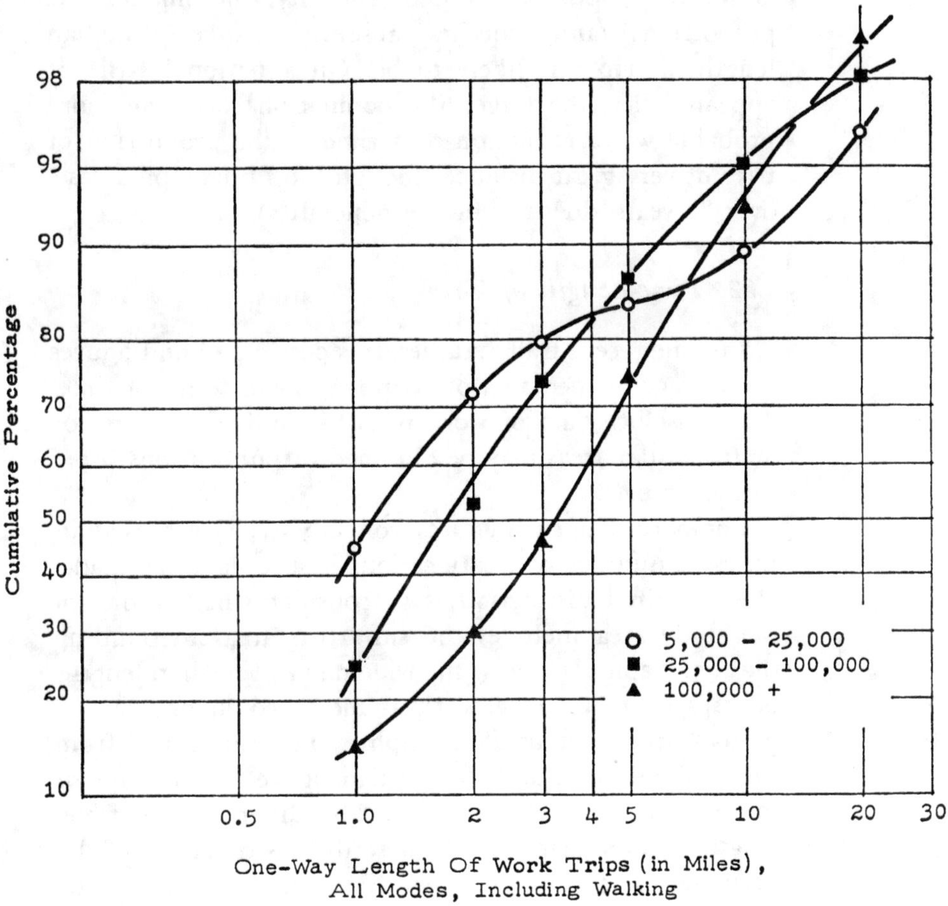

To summarize, it appears that work trips of moderately long distance increased in length from 1941 to 1951 in cities of the several size groups considered. However, the larger the population size of the city, the smaller the proportional (and sometimes absolute) change in median length of trip was likely to be. On a national basis, it appeared that the larger the population class, the more probable was a reduction over time in the proportion of trips of very great distance (though the Philadelphia case, over 22 years, does not fit this generality).

2.22. *Time-Length of Trips*

In some recent traffic studies, it is possible to find figures for average or median work-trip lengths in terms of time. Time-distance data of work trips for individual cities or metropolitan areas may be obtained in publications listed in footnote 4.

Comparison of such figures from city to city is hazardous unless a number of qualifications of the data are made. One required distinction is to consider whether or not the study area includes the suburban area surrounding the city. Table II-8 shows the wide range of median lengths of trips from home to work as measured in one metropolitan area, that of Philadelphia. The range was from 17 to 47 minutes, and the median figure for all trips of interest was about 27 minutes. The differences in time-lengths of these trips may be described as functions of the permutations possible between living and working in the central city or suburbs.

[4] Homer Hoyt and L. Deward Badgely, *The Housing Demand of Workers in Manhattan*. New York, 1939; London Transport Executive, *London Travel Survey*, 1948; South Side Planning Board, *Employee Transportation Survey*. 1952. Chicago, Ill.; Detroit Metropolitan Area Regional Planning Commission, *Home Location Patterns of Industrial Workers in the Detroit Region*. December, 1955.

TABLE II-8

PERCENTAGE DISTRIBUTIONS OF DURATION OF WORK TRIPS OF EMPLOYED HOUSEHOLD HEADS PHILADELPHIA AND THE PSMA, 1956

Time-distance (in minutes)	*Living and working in Phila- delphia*	*Live in Phila- delphia, work in suburbs*	*Live in suburbs, work in Phila- delphia*	*Live and work in suburbs*	*All workers in PSMA*
0- 9	15	2	[a]	30	17
10-19	20	5	5	28	20
20-29	17	7	13	20	18
30-39	21	22	32	14	20
40-49	12	20	23	4	11
50-59	5	9	9	1	4
60+	10	35	18	3	10
Median (in minutes)	28	47	40	17	27
Number of trips	348,100	28,000	125,600	297,400	799,100

Source: U.S. Bureau of the Census, *National Housing Inventory*, Philadelphia Supplement. Data pertain to December, 1956. Field data were obtained by the Bureau of the Census on a service contract with the Institute for Urban Studies, University of Pennsylvania, and were used primarily in connection with housing market analysis. Data have been made available by Dr. Chester Rapkin of the Institute for Urban Studies.

[a] Less than one per cent.

2.23. *Annual Income and Time-Length of Work Trips*

In the Philadelphia Standard Metropolitan Area (eight counties, of which the coterminous county and city of Philadelphia comprises the central area) in 1956, it was found that shorter work trips are "privileges" of the low- and relatively high-income categories of income-earners (Table II-9). Longest trips (33 minutes) were those taken by heads of households in the $7,000 to $7,999 bracket. The greater the difference in income from this amount, either below or above it, the shorter the number of minutes of the median trip. Shortest trips were those of the

TABLE II-9

MEDIAN TIME-LENGTH OF ONE-WAY WORK TRIP IN RELATION TO INCOME OF HOUSEHOLD HEAD PHILADELPHIA AND THE PSMA, 1956

	Length of work trip	
Annual income	*City of Philadelphia* (*to nearest minute*)	*PSMA*[a]
Less than $2,000	28	25
$2,000 to $3,999	30	26
$4,000 to $4,999	33	26
$5,000 to $5,999	33	28
$6,000 to $6,999	30	29
$7,000 to $7,999	32	33
$8,000 to $9,999	38	32
Over $10,000	18	26
TOTAL	30	27

Source: U.S. Bureau of the Census, *National Housing Inventory*. Philadelphia Supplement, 1956.

[a] Philadelphia Standard Metropolitan Area. Bucks, Chester, Delaware, Montgomery, and Philadelphia Counties in Pennsylvania; and Burlington, Camden, and Gloucester Counties in New Jersey.

household heads who earned less than $2,000, but the median of their trips was less than one minute different in length from that of the household heads who earned over $10,000.

The pattern for the City of Philadelphia is by no means so regular as that of the Philadelphia SMA. Still, the extreme income groups have the shortest trips, though of the two, the higher-income extreme has much the shorter median. The longest trip in this array is found for the income group in the range $8,000 to $9,999. In comparing time-distance data of the city with those of the eight-county PSMA, it must be remembered that the Philadelphia data are contained within the metropolitan totals, so that trips for household heads resident in the other seven counties of the PSMA were even more different in length from trips of Philadelphians than the medians for the two distributions would seem to indicate. Comparable data on variations in trip length with income for other urban areas are not now available so it is not known how general may be the patterns reported here.

2.3. MODE OF TRAVEL

The mode of travel selected by an individual is a personal decision made upon such bases as the actual condition of available facilities, the relative time and costs of travel, relative convenience (including waiting times, connections, and walking distances), and the less easily measured elements of how much the traveler knows about the choices open to him, and how he perceives the situation. For individuals, the choice is important but not crucial. For cities, the choices made by large numbers of people have become crucial in recent years. The long-term trend over the years toward use of the personal automobile and away from use of public transit, for example, is directly related to the increasing demands placed upon the street, highway, and parking facilities of the cities. (This matter is discussed more fully in Chapter V.) This same trend has an effect, eventually, upon freedom of choice in sites for housing, commercial, and industrial purposes. Since capacity requirements of both highway and transit facilities are conditioned by peak-period demands, and thus by the demands primarily of work-trip traffic, the modes employed in work trips are of vital concern.

Table II-10 indicates the way in which usage of the vari-

TABLE II-10

PERCENTAGE DISTRIBUTION OF MODES USED IN TRAVELING FROM HOME TO WORK, BY POPULATION OF CITY
SIX STATES, 1951

	Population size group		
	5,000 to 25,000	*25,000 to 100,000*	*100,000 plus*
Passenger car	64.4	62.8	46.4
Public transit	4.2	15.7	38.4
Passenger car and public transit	0.9	0.9	2.2
Walk	24.2	17.2	9.9
All other means and not reported	6.3	3.4	
TOTAL	100.0	100.0	100.0

Source: T. A. Bostick, R. T. Messer, and C. A. Steele, *op. cit.*

ous modes is distributed in the journey to work as measured for a number of cities of varying size in 1951. The dominance of the automobile can be seen in these figures, though the proportion of auto usage is seen to decrease with increasing city population size, where public transit service is more widely available. However, the city size cannot be considered the sole conditioning variable. It is evident from Table II-5 that the length of trip to be taken was at least a joint variable. The figures indicate that residents of the cities of population 25,000 to 100,000 rarely selected transit for work trips of over five miles in length, but rather selected auto or auto-plus-transit for more than 80 per cent of these longer trips.

The story was different in the cities of over 100,000 population. In these, for work trips of under five miles, less than half the residents used autos. But on the trips to work exceeding five miles in one-way length, well over 60 per cent of these residents either drove or rode in automobiles.

On the subject of length of transit trip by size of city, consider Tables II-6 and II-7. As expected, the median lengths increased regularly (from 1.9 to 3.28 miles) with city population (and generally, therefore, with increasing extent of urban area). But it was surprising to find that the *average* length of transit trips was fully 1.5 miles longer for cities of 5,000 to 25,000 population than it was for cities in the next larger group (25,000 to 100,000), and one third of a mile longer than the trips to work in the largest group of cities studied. Thus, it must be inferred that of the relatively few who used transit to go to work in the small cities a large proportion took very long trips. This matter is further discussed under the topic of open-country commuting in Section 4.12.

The figures of Table II-11 document the qualitative judgment that in the last generation a completely new kind of urban travel picture has come into being. Changes are presented in selection of modes for home-to-work trips in a major city, measured over a 22-year span. The change in relative transit and auto usage is amazing. The reduction in proportion of walkers from 1934 to 1956 is also to be noticed. The drop from about 23 per cent to 9 per cent is significant, even when it is considered that the data are the results of sampling surveys.

Chapter V details considerably more information on changes in modes of urban travel over recent years.

2.31. Spatial Pattern in Selection of Mode

The evidence available indicates that location within the metropolitan area is a strong determinant of choice of mode used in urban travel. Similarly, as discussed in

TABLE II-11

PERCENTAGE DISTRIBUTION OF MODES USED IN HOME-TO-WORK TRIPS PHILADELPHIA RESIDENTS, 1934 AND 1956

Mode	*1934*[a] *Main wage earner*	*1956*[b] *Head of household*
Auto driver and passenger	12	50
Public transit	57	35
Railroad	1	1
Combined and other modes	[c]	1
Walk	23	9
Work at home	7	4
TOTAL	100	100

[a] *Report of Philadelphia Real Property Survey, 1934.* Summary, Residential Structure and Family Accommodations, Federal WPA for Pennsylvania, Project 4744. Total trips reported above = 345,531, excludes 81,796 families with no wage earner and another 29,291 which gave no information. Possible total was 456,618.

[b] From Alderson and Sessions, *The Families of the Delaware Valley.* December, 1956. Excludes 104,000 heads of households who were not employed, because they were unemployed, retired, with an independent income, etc.

[c] Not reported.

Chapter III, measurable factors which tend to be functions of location (e.g., net residential density, distance from city center, income, and auto ownership) are to varying degrees associated with the numbers of daily trips generated from each dwelling unit.

A report on trips to the Philadelphia CBD presents the inference that zonal patterns in the selection of mode tend to prevail throughout the range of trips for all purposes.[5] Where a single mode particularly characterized a section of the city, it was found that that mode tended to predominate throughout all the trip-purpose categories. Since the *purpose* of an urban person-trip does not appear as a primary factor in selection of mode, our key to the study of mode selected for trips originating from home is more likely to be found on an area basis. Area characteristics considered to be relevant to some degree in study of modes selected are per capita auto ownership, income levels, occupation of residents, net residential densities, and location of residence relative to the major centers of destination.

For the present effort, a geographic pattern of selection of public transit (surface transit plus subway-elevated lines) for work trips from internal-area origins was studied from the specially tabulated O-D data for Philadelphia (1947). Because the use of transit has declined considerably relative to private automobile transportation in the journey to work since the time of the Philadelphia area 1947 O-D survey, the absolute figures presented are not valuable. However, analysis and presentation of such data not only can provide information against which more recent data may be compared, but more importantly, can indicate pattern and explain that pattern in terms of independently measured demographic, social, or physical factors. Hence, the sifting and assay of 1947 data are presented with qualification but without apology.

All work trips originating within the Philadelphia transportation survey area, except for some districts in New Jersey, were grouped according to mode selected: that is, by private vehicle or by public transit. (Taxi trips were grouped with automobile trips.) The proportion of work-

[5] From information on the 1947 O-D study developed for use in the report, *Person-trips to the Central City*. Philadelphia City Planning Commission, 1953. Developed by Hans Blumenfeld and Frederick Sass.

ers selecting public transit was computed for each originating area, and contours of equal percentage were drawn.

When plotted as percentage contours (a Weberian term such as "isomode" might be coined to describe the lines), the data appeared consistent in area patterns around the central focus of the metropolitan area. The patterns appeared to corroborate the hypothesis that choice of mode is closely associated with facets of spatial location of residence.

In the study generally, it was found that residential areas of relatively high transit usage on work trips were located radially from the center city along old well-established transit lines. Transit usage was found to decay regularly with decreasing density throughout the area, but along the radial high-speed lines fingers of a slower rate of decay were noted.[6]

Secondly, as was expected, it was found that high usage of transit for work travel occurred most often in middle- and low-income areas with quite high population density. That this geographic distribution of modal selection was fairly regular is not surprising, since for trips of all purposes throughout the city, the proportion of riders who selected transit over auto was found to vary regularly from 77 per cent for families paying less than $30 per month rent or its equivalent, to 38 per cent for those who paid over $80 per month.

2.32. Effects of Auto Registration and Associated Location-Sensitive Variables

Selection of mode, not surprisingly, is closely associated with levels of auto ownership. The point may be demonstrated in these two ways: by considering the relationship

[6] The term decay is used here to indicate the decrease in an ordinate value of a curve, with increasing abscissa value. The term derives from the physical science usage, e.g., the decay of an existing electromotive force in a conductor, over time, after interruption of a previously closed circuit.

between the persons-to-autos ratio and the relative proportions of auto and transit usage in each of a number of small districts within an urban area at a single point in time; and by a time-series analysis of per capita transit rides per year with the annual average persons-to-autos ratio for the transit service area. Both types of analysis have been undertaken, using Philadelphia data. Patterns observed in each case were found to be clear and coherent. The second type of analysis, the time-series approach, is described in Chapter V.

In the first-named type of study, that which cross-sections the area at a single point in time, a graphic multiple correlation study was performed, using the percentage of work trips by transit mode from each area as the dependent variable.[7] Associated factors were net residential density, persons-per-auto ratio, and distance of residential area centroids from the center of the CBD. Median family income was not included, though it would have been reasonable to introduce this factor into the study.

The principal value of the graphic study with the 1947 data is in the finding of clarity of pattern, using the variables listed. It seems archaic to discuss an observed value of 80 per cent transit riding on work trips (for areas of about 180 persons per acre net residential density, or areas with persons-per-auto levels of about 18). The densities have not changed greatly in intervening years, but the levels of transit riding and auto registration certainly have.

In general, as may be seen from data in Table II-12, areas of highest transit usage were those of highest density, shortest distance from center of the CBD, and highest persons-per-auto ratios. Data are given for the 12 planning

[7] The Bean or Ezekiel method was employed. See Chapter 16 on short-cut methods of multiple correlation, Mordecai Ezekiel, *Methods of Correlation Analysis* (2d ed., New York: John Wiley and Sons, 1941).

TABLE II-12

PROPORTIONAL USE OF PUBLIC TRANSIT FOR ORIGINATING VEHICULAR WORK TRIPS,[a] AND FACTORS ASSOCIATED WITH SELECTION OF MODE

PHILADELPHIA, 1947

Planning analysis section	Transit work trips: Per cent of section total work trips	Transit work trips: Per cent of city transit work trips	Persons per registered auto	Net residential density, persons per acre, 1950	Median income per family, 1949 (in dollars)	Transit trips per dwelling unit, all purposes[b]	Distance residence centroid from city center (in miles)
A	86.0	2.2	12.4	170.1	1,723	1.58	0.4
B	83.6	15.0	18.1	191.7	2,512	2.89	2.0
C	61.7	5.2	9.7	106.2	3,395	3.44	3.6
D	64.6	16.2	8.9	113.0	2,721	3.53	3.4
E	80.0	19.9	18.5	202.2	2,040	3.22	2.2
F	71.0	8.5	10.1	131.1	3,080	3.83	4.2
G	67.8	7.0	12.6	146.2	2,718	4.22	4.4
H	50.0	1.6	8.0	42.1	3,206	3.14	6.8
I	51.0	6.5	5.9	36.0	3,464	4.38	6.6
J	42.7	9.0	6.2	80.4	4,894	5.08	6.4
K	46.5	8.6	6.2	73.0	3,810	4.03	7.6
L	30.4	0.3	5.0	23.5	3,398	3.19	12.2
City	63.6	100.0	9.8	100.4	2,869	3.58	

[a] Traffic data are for average weekday, June to November, 1947.
[b] Includes taxi trips also.

analysis sections of Philadelphia, but equivalent measures have been prepared for the 67 component subsections, from work by Blumenfeld and by the writer. At present it seems sufficient to note the consistency of findings using the short-cut graphic methods. At such time as data from the forthcoming O-D survey of the Penn-Jersey Transportation Study become available, it will probably be profitable to develop further the measures of association of a number of the micro-area variables, using 1947 data as well as those of 1960.

Mentioned above is the ratio of persons to autos. This ratio is an important factor in present micro-area studies of trip generation and selection of mode. Hence it is worth noting that differences between figures for vehicle registration and for vehicle ownership can be misleading. Because title may change during the year or more than one registered owner in a family may result in multiple registrations for the same vehicle, the registration figures may exceed those of ownership by a proportion that has sometimes been measured at about 10 per cent. Data obtained from residential interviews should preferably be directed to yield ownership information for vehicles used in urban transportation service. It should be noticed also that the persons-per-auto ratios are dependent for their usefulness on the uniformity of classification of "home-base" or primary garaging location of the individual vehicles.

2.4. TIME CHARACTERISTICS

Not a great deal is known of the seasonal, monthly, or weekly patterns of the journey to work.[8] However, in terms of the hourly distributions of work trips in the average day studied, the peaking characteristics are known to the point of notoriety. Work travelers themselves may be poor accountants of the time and costs of their trips on a cumulative basis, but they are subjectively well aware of the crush hour.

In considering friction in the elements of a city's transportation plant, it might be noted that psychological friction exists as well as actual friction and that, in effects on actual travel patterns, one type is as real as the other. This point may be illustrated by the response of transit riders to nonstop or "limited" services. The psychic effect of the stops and delays on the nonlimited service is out of proportion to the few minutes of time lost. Or, consid-

[8] Hopefully, this statement may be proved untrue before much more time elapses. The Port of New York Authority has initiated procedures for an annual evaluation of O-D materials available in the PNYA service area. Refer to Lovejoy's article in Highway Research Board *Bulletin 224*.

ered quantitatively, mean speeds do not register with the individual as do a number of repeated accelerations and decelerations. It is believed that physical barriers and traffic obstacles in the metropolitan terrain, such as noticeable changes in grade and elevation and, most of all, the central business district, in similar fashion create impressions of friction that are out of proportion to actuality. This is not to discount real friction and real stops and delays or the real separation caused by a river or lake, but to point out that apparent distance and apparent friction produce final results that are measurable and mappable.

To gain a quantitative view of the peaking characteristic the reader may refer to Figure C-1, Appendix C, which shows the hourly volumes of trips for all purposes in one O-D survey, as measured at two screen-line stations. The work-travel proportion of the traffic enumerated during the hours of highest movement volumes, it should be remembered, may well exceed three fourths of the total. In Detroit, for example, in 1953, it was found that 90 per cent of the morning peak-period traffic and 78 per cent of the evening peak-period was composed of vehicles en route to or from workplaces.[9]

Traffic measurements made in Minneapolis and St. Paul (September, October, 1949) indicated considerable differences between average speeds attained on urban facilities, as measured during off-peak and peak periods of traffic flow (Table II-13).

The St. Paul data indicate that about 45 per cent more distance was covered in the first 15 minutes of travel from the downtown area in the off-peak as compared with the peak period of traffic flow. The equivalent figure for Minneapolis was 18 per cent. Despite the considerable difference between these two figures, the retarding effect

[9] Detroit City Planning Commission, *op. cit.*

TABLE II-13

Comparison of Average Distances from Central Business District Traveled in Peak and Off-Peak Periods

Minneapolis and St. Paul, 1949

	Cumulative distance			
	Minneapolis		*St. Paul*	
Cumulative time (in minutes)	*Peak*	*Off-peak*	*Peak*	*Off-peak*
	(in miles)			
5	0.8	1.0	0.6	1.1
10	2.2	2.6	2.2	3.2
15	3.9	4.6	4.4	6.4

Source: Minnesota Department of Highways, *St. Paul-Minneapolis Traffic Survey*, 1950, pp. 43-45.

on traffic of the great volumes of workers who travel at the same time is evident.

2.41. Time of Arrival: Work Trips in Four Urban Areas

Time of arrival patterns for work trips in four urban areas were studied, particularly with reference to economic character as revealed by the industrial composition of employment. Of the four surveys, the earliest was Philadelphia-Camden, taken in 1947, and the next was Erie, Pa., in 1948. Field data for Madison, Wis. and Minneapolis-St. Paul were taken in 1949. Although the sample of four cities is too small to warrant generalization, the results obtained for the areas studied are of interest. One application of the type of information gathered would be in analysis of O-D materials for design of a staggered-hours program.

The patterns in the time of arrival for the two largest of the four areas may be seen to be quite similar (Table II-14). The two smallest, Erie and Madison, were found to be notably divergent. The industrial composition of the Minnesota urban area, incorporating Minneapolis and St. Paul, was similar to that of Philadelphia-Camden in 1950 (Tables C-8 and C-9). However, the relative strengths of manufacturing versus wholesale and retail trade were reversed in the two areas. Whereas Philadelphia-Camden was strongest in manufacturing, Minneapolis-St. Paul had the highest proportion of local employment in wholesale and retail trade. Madison had a fairly low manufacturing proportion of employment, but was high in the trade, government, and professional categories. The latter two are related to the location of the State capital and university in Madison. Though the employment composition of the Erie urbanized area is not given, it is clearly a manufacturing center, with 48.4 per cent of its 61,627 employed (1950) engaged in manufacturing industry.

Not much can be said for the time patterns of the two smallest urban areas, other than to note that the proportions of trips arriving between 7 a.m. and 9:29 a.m. were lower in these than in the two larger areas. It should also be observed that nothing is known of the time patterns of over 10 per cent of the work trips in both of the smaller urban areas, hence the precision of the proportions is probably low. One curious trend may be observed in the table, although it may only be accidental: that for these four areas the proportion of trips arriving in the period 12 noon to 3:59 p.m., the second largest time grouping, was inversely proportional to the population size of the urban area. On the other hand, of course, the proportion arriving in the largest time grouping, from 7 a.m. to 9:29 a.m., increased with the population size of urban area (though the percentage difference observed for this grouping be-

TABLE II-14

Percentage Distributions of Work Trips by Time of Arrival at Destinations Internal Surveys, Four Urban Areas[a]

Time period	Erie, Pa., 1948		Madison, Wis., 1949		Minn.-St. Paul, 1949		Phila.-Camden, 1947	
	Number	Per cent	Number	Per cent	Number	Per cent	Number	Per cent
7:00 a.m.- 9:29 a.m.	28,725	52.3	15,426	40.3	226,376	58.6	552,152	59.2
9:30 a.m.-11:59 a.m.	4,991	9.1	4,842	12.6	35,818	9.3	100,071	10.7
12:00 - 3:59 p.m.	9,640	17.5	8,358	21.8	58,068	15.0	125,727	13.5
4:00 p.m.- 8:59 p.m.	5,719	10.4	4,802	12.5	30,782	8.0	69,512	7.4
9:00 p.m.- 6:59 a.m.	5,894	10.7	4,919	12.8	35,053	9.1	85,623	9.2
Subtotal	54,969	100.0	38,347	100.0			933,085	100.0
Unknown time	5,995		4,281				1,411	
TOTAL	60,964		42,628		386,097	100.0	934,496	

[a] Sources and notes: Distributions for Erie, Madison, and Minneapolis-St. Paul calculated from data made available through courtesy of the Highway Transport Research Branch, U.S. Bureau of Public Roads.

Distribution for Philadelphia-Camden internal area calculated from materials made available through courtesy of Philadelphia City Planning Commission. Philadelphia-Camden data includes 928,993 trips to destinations in districts 000 through 474, and 5,503 trips to destinations in area-units 508 through 730.

tween the two largest was not statistically significant). How general this trend by population size may be is not known.

2.42. *Time of Arrival: Work Trips to CBD*

The central business district represents a major, if not the largest, locus of work-trip destinations in urban areas. To gain some understanding of time-of-arrival characteristics in a CBD, time distributions were prepared for the 10 component areas of the Philadelphia CBD (Table II-15).

Although there were minor differences, as noted below, the most notable characteristic of these distributions was their similarity, indicating that the component subareas of the CBD function together as a monolithic trip-generator insofar as traffic-peaking characteristics are concerned. Another striking point is the difference between time distributions for work trips and for trips for all purposes to the CBD as a whole; 40.4 per cent of the latter arrived in the 2½-hour period 7 to 9:29 a.m. while in the same period 68.5 per cent of work trips to the CBD reached their destinations. Arrivals of trips for all purposes tended to be fairly evenly distributed throughout the remainder of the day, while the proportion of CBD-destined work trips dropped off much more quickly after the morning peak.

This pattern, however, did not hold for all component areas of the CBD. In three, the proportions of work trips remained relatively high, at about 15 per cent of the total in the late morning hours, and in one district, 002, this figure of 15 per cent of the daily work trips held into the afternoon period. The three traffic districts with the characteristic of late-arriving work trips are contiguous, comprising the area within which are located the major department stores, the large banking, finance, and insurance establishments, and enterprises ancillary to major

TABLE II-15

WORK TRIPS ARRIVING IN CBD FROM ORIGINS[a] IN THE INTERNAL SURVEY AREA BY ALL VEHICULAR MODES, BY DISTRICT AND TIME OF ARRIVAL

PHILADELPHIA, 1947

Trip purpose	*Destination district within CBD*	*Ratio:[b] work trip to total trip destinations*	*Percentage of trips by times of arrival*						
			9:p.m. to 6:59 a.m.	*7:a.m. to 9:29 a.m.*	*9:30 to 11:59 a.m.*	*Noon to 3:59 p.m.*	*4:p.m. to 8:59 p.m.*	*No answer*	*Total*
Work	000	0.50	5.22	67.80	14.75	7.43	4.55	0.25	100.00
	001	0.83	8.20	72.12	4.14	8.96	5.09	1.49	100.00
	002	0.41	4.27	56.11	15.05	14.68	9.08	0.81	100.00
	003	0.57	5.88	72.04	8.19	6.42	7.47	0	100.00
	004	0.43	5.29	62.67	15.80	8.28	6.55	1.41	100.00
	005	0.72	13.77	65.20	7.66	9.39	3.27	0.71	100.00
	006	0.67	3.42	73.57	9.00	9.31	4.09	0.61	100.00
	007	0.85	3.92	82.21	5.02	3.63	4.79	0.43	100.00
	008	0.78	5.04	72.17	7.45	9.25	5.69	0.40	100.00
	009	0.85	4.78	76.65	7.48	5.57	5.25	0.27	100.00
	00(CBD)	0.55	5.89	68.46	12.14	8.10	4.97	0.44	100.00
ALL PURPOSES	00(CBD)	—	7.38	40.44	17.40	17.38	17.33	0.07	100.00

Sources: Special tabulations of work-trip data prepared for the present study, from cards of *Philadelphia-Camden Area Traffic Survey*, Pennsylvania Department of Highways, *et al.*, 1950. Data for trips of all purposes from *Person-trips to the Central City*, Philadelphia City Planning Commission, 1954, p. 90.

[a] Residence within the limits of the outer cordon.

[b] Calculated for internal trips entering CBD, but excludes from total the trips of purposes "change travel mode" and "serve passenger."

trade and finance. There is a considerable amount of personal service employment, such as that in medical offices, legal offices, and other establishments of a professional nature. District 002, in particular, with its high proportion of late arrivals for work, is a center for private offices of physicians and other medical personnel. The personal service nature of these three districts is also indicated by the fact that they are the only ones in which work-trip terminations comprise less than half of the vehicular trip terminations for all purposes (see Column 2 and footnote *b* in Table II-15).

District 005, with its high proportion of early-hour trips, has served as a major goods distribution and industrial area. In general, the data indicated that work trips to wholesaling (particularly food wholesaling) areas may be expected to peak earlier than those to trade and finance areas. Secondly, the proportions of trips to loci of personal services may be expected to peak less sharply in the morning and to hold up through the day at a level above average for the CBD. The net effect in the central business districts of most cities of the trips for varying purposes, with their individual patterns of hourly distribution of arrivals and departures, is realized in a maximum accumulation of persons and vehicles in the hours from 12 noon to 2 p.m.

Information on the hourly and directional distributions of trips by specific routes is used in highway planning. The peak-hour multiplier, or proportion of the peak-hour traffic to average daily traffic on a given route, is an important factor in determining the capacity of a highway facility. As more knowledge is gained of time-patterns of travel, by origin and destination areas and by purposes of trips, the more accurately it should become possible to predict the critical multiplier factor for vehicular routes. Further, this kind of information, as it becomes known for more cities and can be generalized in terms of trip-generation characteristics, should be of use in connection with future efforts to reduce peak volumes by staggering work hours and widening the hourly plateaus of traffic volumes.

2.5. PERSONS PER VEHICLE AND RIDE SHARING

Work trips impose the most intense demands upon the capacity of local circulation facilities because of the great numbers of persons- and vehicle-trips made within limited time periods. But while the public transit vehicles carry their greatest loads during the periods of work-trip dominance, auto occupancy for work trips is lower than for any other of the major systems of trips.

The point made above may be demonstrated from data of two studies made in 1948 (Table II-16). On a traffic cordon line drawn in the San Francisco Bay Area at the periphery of the most densely settled part of the metropolitan area, the average occupancy on work trips was found to be 1.5 in contrast to an over-all average of 1.9. Similarly, for the work trips measured in the internal area of the Sacramento O-D survey, occupancy of passenger cars was 1.6, lowest of any other trip purpose, as opposed to the average for trips of all purposes of 1.9.

Studies made by the Pennsylvania Deartment of Highways at cordon stations around the Philadelphia-Camden survey area in 1955 show a typical pattern of lower auto occupancy on work trips in comparison with trips for all purposes (Table II-17).

Some generalized figures on occupancy have been presented in *Traffic Engineering Handbook,*[10] summarized from O-D surveys conducted from 1945 to 1948 in 13 cities (Table II-18).

[10] Institute of Traffic Engineers, 1950, edited by Henry K. Evans, pp. 44-45.

TABLE II-16

AVERAGE NUMBER OF PERSONS PER AUTO BY PURPOSE OF TRIP
SACRAMENTO INTERNAL AREA AND
SAN FRANCISCO BAY AREA CORDON LINE, 1948[a]

Trip purpose	*Average passenger car occupancy, 1948*	
	Sacramento, internal area	*Bay Area, cordon line*
Work	1.6	1.5
Transact business	1.7	1.7
Medical, dental	2.2	2.2
School	2.0	1.8
Social, recreation	2.4	2.4
Eat meal	2.0	2.4
Shopping	2.2	2.3
Change travel mode	2.4	1.9
Serve passenger	2.3	2.3
Home	1.9	1.8
Unknown	2.0	2.0
OVER-ALL AVERAGE	1.9	1.9

[a] Source: California Division of Highways.

TABLE II-17

AUTO OCCUPANCY BY PURPOSE OF TRIP
PHILADELPHIA-CAMDEN CORDON STATIONS, 1955

Purpose	*Average auto occupancy*		
	Local	*Through*	*All*
Work	1.4	1.4	1.4
Business	1.6	1.6	1.6
All	1.8	2.2	1.8

TABLE II-18

MEDIAN AUTO OCCUPANCY BY TYPE OF TRIP, 13 CITIES, 1945-1948

Trip classification	*Auto occupancy*
All internal	1.5
External	
out-of-town[a]	1.9
through[b]	2.3
all	1.9
All	1.5

[a] One trip-end in internal area.
[b] No trip-end in internal area.

It can be seen that trips crossing the cordons around the urban areas are likely to have higher vehicle occupancies than internal trips. While some of these cordon-crossing trips might be quite short, it is a fair assumption that most trips with one or both trip-ends outside the internal area are of longer length than those made wholly in the internal area. An inference from this would be that vehicle occupancy tends to increase with length of trip. A finding corroborating the inference was made from traffic data for Flint, Michigan, where it was found that ride sharing in autos tended to increase, up to a trip length of about 30 miles. Beyond this limiting distance, occupancy tended to remain constant.[11]

2.6. SUMMARY

Work trips, as measured in internal O-D surveys, represent from one sixth to one third of urban vehicular trips.

[11] Schnore, Leo F., *The Separation of Home and Work in Flint, Michigan* (Ann Arbor: Institute for Human Adjustment, University of Michigan, 1954).

The internal area journey to work, therefore, comprises a range of from less than one third to less than two thirds of vehicular trips.

Care must be taken in reporting such proportions to specify whether the total group of trips includes those in the relatively less usable categories, to home, to change travel mode, and to serve passengers, or whether it includes only those to nonhome ultimate purposes.

Length of work trip is found to be a function of the size of the city in which the worker resides and of location within the urban area. The changes in work-trip length in a recent decade were also found to be related to the population size of the city of residence.

The mode of travel selected for work trips is found to vary with differences in city size, with the spatial location of the residence of the traveler, and with a number of associated factors measurable in each of the subareas within the survey area. These factors are relevant to all aspects of trip generation; an instance is the interrelationship between the mode selected in a given size of city and the length of work trip to be taken.

The peaking characteristic of work trips during the day is a fact of life in cities. This characteristic imposes stresses upon the travelers and physical loads upon the transportation plant that demand further study. One major type of study effort, responsive to this problem area, calls for a structuring of the destination trip-ends for work travel in terms of the respective trip origins and times of arrival. The so-called staggered hours studies illustrate one kind of effort directed to this difficult situation.

The number of persons in private vehicles used in the journey to work appears to average only about three fourths of those in auto trips for all purposes. The low load factor intensifies the difficulty of providing adequate capacity of vehicular facilities in periods of peak demand.

This matter should be of concern not only to traffic and planning officials; it relates directly to the individuals concerned and to their employing establishments. Incentives for increasing participation in car-pools in the journey to work appear clearly desirable, particularly for those in-city establishments where parking space is in high demand but in critically short supply.

III

Work-Trip Origins: Persons and Places

CHARACTERISTICS OF THE TRAFFIC OR MOVEMENT ASPECTS of work trips have been discussed in the previous chapter. These represent but a single way of looking at the systems of work travel. Other major viewpoints useful for analytical purposes include the following:

1. the person who makes the trip
2. the place of origin of the trip
3. the place of destination of the trip
4. the relationship between origin and destination (home and workplace)

Of these four, the first represents consideration of the actor in the situation. The other three concern the environment and the interrelationships of the parts of the trip. Synthesis of information from these several viewpoints can lead to the finding of pattern, and therefore to development of tools of prediction of future travel behavior in the journey to work.

This chapter considers the journey to work in terms of the first two of the four viewpoints listed above. The remaining two are discussed in Chapter IV. Illustrations are given of studies of work trips from residential zones within a city. Travel patterns and characteristics of cities are considered, as are the potential uses of information on daytime distribution of population. The essence of the present effort is the consideration of component parts of the journey to work and their interrelationships, the interaction between the journey to work and its environment, and illustrations of means of pertinent analyses.

3.1. CONSIDERING THE WORK TRAVELER

Given primary emphasis in the present study are those attributes of work travelers which aid in understanding the size, distance, directional and time patterns of the streams of movement, with the people in those streams considered en masse. This emphasis includes the elements of human motivation to the extent that these are implicit in the observed phenomena. The attributes considered are those which may readily and economically be measured by the methods of governmentally sponsored sampling surveys. This kind of information has been used in surveying the market for requirements in public transportation facilities and in planning for new construction, but the summary data available cannot be considered to indicate the innermost reasons guiding the travel behavior of the human beings for whom the facilities are to be provided. This known limitation on the information, however, should not lead to disparagement of the difficult-to-obtain data that we do have. The justifiable resistance of individuals against governmental intrusion into their personal lives is a real boundary line for official investigations into personal motivations. We shall have to continue to develop techniques of analysis, inference, and planning and to expect that future basic data inputs shall essentially be variations on presently known themes.

The present data tell us who is likely to travel in cities,

and if we analyze them carefully, they tell by what mode and to where the trip is made. The variation needed is the "why." Further application of the powers of the sciences of the humanities to these problems would be most welcome. Sociological aspects of travel behavior have been considered in the past; however, the powers of the psychological arts and sciences have yet to be applied to many parts of the present subject.[1]

At present we cannot say which descriptive factors about work travelers (or for that matter, about persons traveling for any purpose) are most valid for predictive purposes. In the last few years, however, consensus has been developing on measures of current relevance to trip generation. How well these measures will fit the needs of traffic prediction is not known. Nonetheless, first steps have to be taken if a tool of any power is to be developed. In the absence of better information, we are developing the tools to predict future travel patterns upon the premise that factors which measurably affect present travel behavior will similarly affect travel in the future.

In the effort to understand recent travel patterns, some of the attributes of the traveler that have been studied include those which indicate place in the family and in the labor force, age, sex, family income, and access to vehicular transportation. Knowledge of whether or not a person is head of his household, or primary or secondary wage-earner, is important for interpretation. The family-position categories tend to incorporate information about age and sex of the traveler, which in the past have been found associated with some travel characteristics. As might be supposed, but as studies confirm, access to vehicular transportation can be called a key factor. In the traffic

[1] Aspects of possible working relationships among technical disciplines were described by Mrs. M. E. Meyerson in her paper given in Washington, D.C., before the Highway Research Board, January, 1955: "Implications of Sociological Research for Urban Passenger Transportation Policy."

analysis it is reasonable to consider family income levels since production of income normally requires some amount of local travel and the income obtained affects the degree of freedom that members have to pursue objectives and interests in the community. Factors appearing correlative with income level generally have been found to have a significant association with trip generation and travel patterns. Such factors include monthly rental or, for home owners, imputed monthly rental, occupation, and automobile ownership. It may be seen that data on area distribution of automobile ownership, *at present,* tend to incorporate information on income level as well as that on access to vehicular transportation. As might be expected from this dual nature, the index of automobile ownership has been found useful for present purposes of travel prediction.

There is no assurance that the factors named here are the most basic or important or stable over time for describing or predicting travel behavior in cities. It can be said that they have present or recent association with measured travel patterns. One certainty at this stage of development in the art and science of prediction of travel behavior is that very little is to be dismissed as "obvious."

3.11. Characteristics of the Labor Force

The volume of the journey to work relative to all urban traffic is conditioned by the proportion of the urban labor force to the total population. Let us consider this proportion in its elemental aspects.

The labor force is held generally to be the employable persons of 14 years of age or more, whether they are presently employed or are seeking work.[2] From the viewpoint of the present effort, we are most concerned with that part of the civilian labor force which is gainfully employed.

[2] See Glossary for more detailed description of terms used.

This includes managers, officials, and proprietors, not only wage and salary earners.

The scale of the labor force and of its elements relative to total population is shown below, for the total United States population, the total urban population, and, because a number of traffic examples are given for that city, for Philadelphia (Table III-1).

TABLE III-1

EMPLOYED PERSONS RELATIVE TO TOTAL POPULATION
UNITED STATES, URBAN UNITED STATES,
AND PHILADELPHIA, 1950

	Per cent of total population		
	U.S.	*Urban U.S.*	*Philadelphia*
Persons 14 years old and over	74.6	77.0	78.7
Labor force	39.9	42.6	43.0
Civilian labor force	39.2	42.1	42.7
Employed	37.3	39.8	39.8

The proportion shown for employed persons relative to total population in urban areas, roughly 40 per cent, prevails in most American cities in the succession of post-World War II years that have seen a moderately high level of economic activity. However, because of smaller family size in the more central parts of some metropolitan areas, the proportion may be expected to vary somewhat with density within a given metropolitan area. In the City of San Francisco, the densely settled core city of an extended metropolitan area, the resident employed force in 1950 was 42.6 per cent of resident population, several percentage points higher than the prevailing figure for the entire metropolitan area.

Of special interest to students of the journey to work is the distinction between those employed who work at their residences and those who travel between homes and workplaces. (The original journeymen, it is interesting to note, were those who came in to work by the day, from residences located apart from the workplaces of their craftsmen-employers.) In Philadelphia the postwar work-at-home group is estimated to be 2.5 per cent of the employed force. (In 1934, 7.1 per cent of "main wage earners" were employed at home in Philadelphia.)

The civilian employed force may be broken down into the following Bureau of the Census categories, with particular interest in the last two groups, whose travel characteristics may be expected to differ considerably from those of the total employed (Table III-2).

TABLE III-2

COMPOSITION OF CIVILIAN EMPLOYED FORCE UNITED STATES, URBAN UNITED STATES, AND PHILADELPHIA, 1950

	Per cent civilian employed		
	U.S.	*Urban U.S.*	*Philadelphia*
Private wage and salary workers	71.2	78.9	81.3
Government workers	9.8	10.6	9.2
Self-employed workers	17.0	10.2	9.3
Unpaid family workers	2.0	0.3	0.2
TOTAL	100.0	100.0	100.0

3.12. Primary and Secondary Employment

To the extent that structuring of communities in the aggregate is conditioned by the relationships between residence and workplace of employed individuals, the structuring effect may be attributed for the most part to the characteristics of work trips of employed heads of households. These compose the primary employment, insofar as work-trip patterns are concerned. Other employed members of households compose the secondary employment.

The secondary income-earners are assumed generally to select employment within walking distance or accessible by transit from their homes. This accessibility may be as important in the choice of places of secondary employment as the usual job elements of type of work, hours, vacations and leaves, and fringe benefits. This hypothesis, however, is probably too simple. Where a family knows that more than one of its members will work to provide income, the availability of sources of secondary employment, accessible by walking or transit, might condition to some extent the final choice of a location of its residence. The relative weights of factors in this decision represent types of data that can profitably be analyzed by techniques of multiple correlation. This aspect of the selection of the location of residence, however, is traditionally in the field of housing market analysis, and is not further explored in the present study.

A substantiated hypothesis on urban travel is that the average travel patterns of men and women differ considerably. In Richmond, Va. (1945), for instance, men reported a daily average of 1.27 trips for work or business purposes, but the equivalent average for women was only 0.68. On the other hand, the same study found that the reported daily average for trips for social, recreational, and shopping purposes was 0.45 for women and only 0.25 for men. Another example: in Philadelphia (1947), trips reported by men accounted for 60 per cent of all vehicular trips and 87 per cent of auto-driver trips. Of all the vehicular trips made by women, fully 73 per cent were made by public transit, whereas only 53 per cent of trips by men utilized this mode.[3]

In analyzing the journey to work, it is believed that classifying information on travel patterns in terms of primary and secondary employment incorporates not only the factor of sex of persons making urban trips but also that of age. Thus, this classification is believed to be more useful in the study of work-trip patterns than classification by age and sex.

3.13. Occupation

Occupation of work travelers appears to be related to some features of work trips, though occupation so far has not proved very useful in attempted projections of urban travel. The factor of the occupation of work traveler may in time be found of aid in labor market studies of the area around proposed new industrial sites, to consider distances which might be traveled by the needed labor force under conditions of available transportation facilities. To illustrate some of the differences observed in work-trip features for persons of diverse occupations, some available information for the Philadelphia area is discussed below.

Features of work trips which have been observed to vary with occupation were median time-length of trips (Table III-3), modes selected (Table III-4), and the proportions of total trips made for the major classes of trip purposes (Table III-5).

3.131. Median Time-Length of Trip. This measure, for employed household heads in Philadelphia (1956), was found to vary from 21 minutes for proprietors and managers to 35 minutes for clerical employees. In terms of the median trip length for all occupations (30 minutes), the spread of 14 minutes appears to be very wide. In the eight-

[3] Data of this paragraph from Highway Transport Research Branch, U.S. Bureau of Public Roads.

TABLE III-3

MEDIAN TIME-LENGTH OF HOME-TO-WORK TRIPS BY EMPLOYED HOUSEHOLD HEADS, BY OCCUPATIONAL GROUP PHILADELPHIA CITY AND PHILADELPHIA STANDARD METROPOLITAN AREA, 1956

	Median trip	
Occupational group	*Philadelphia* (*in minutes*)	*PSMA*
Technical and professional	23	28
Proprietors, managers, etc.	21	23
Clerical	35	31
Sales	32	29
Craftsmen, foremen, etc.	33	28
Operatives	30	27
Service	27	24
Other labor	28	28
ALL HOUSEHOLD HEADS[a]	30	27

Source: U.S. Bureau of the Census, *National Housing Inventory*, 1956.
[a] Includes a small number of farm managers and laborers.

county metropolitan area, the spread from lowest to highest median figure is not so great as in the central city, ranging from 23 to 31 minutes. The shortest median trip-lengths for both central city and the SMA were observed for the occupational group of proprietors and managers. This low median travel time may be accounted for in part by a number of proprietors who made their homes immediately adjacent to their places of business. It may also indicate the greater economic freedom of this occupational group, not only in selecting place of residence but possibly also in location of workplace. It may be that after all rational economic considerations in plant location have been duly weighed, the final deciding factor in choosing

TABLE III-4

Percentage Distribution of Modes Employed in Home-to-Work Travel by Gainfully Employed Persons, by Occupational Groups Philadelphia, 1952-1953

Occupational group	*Per cent of total employed*	*Auto driver*	*Auto passenger*	*Train*	*Public transit*	*Auto and public transit*	*Other combination*	*Walk*	*Other modes*	*No travel*	*Total*
Professional, semiprofessional	7.5	38.8	8.4	3.9	33.9	1.0	1.8	2.3	1.0	8.9	100.0
Proprietors, managers	10.3	38.5	4.3	1.3	15.1	0.2	1.0	7.4	1.3	30.9	100.0
Clerks, salesmen	22.1	19.8	6.3	1.5	59.6	0.3	2.1	6.6	0.6	3.2	100.0
Agents, traveling	2.1	68.6	1.9	0.9	19.1	0.9	—	4.8	0.9	2.9	100.0
Craftsmen, foremen	21.9	42.4	8.0	0.8	36.7	0.2	2.1	5.9	1.5	2.4	100.0
Semi- and unskilled labor	24.3	23.0	7.4	0.7	54.7	0.1	1.8	10.0	0.5	1.8	100.0
Protective services[a]	5.0	22.6	1.6	—	17.5	—	—	3.9	2.7	51.7	100.0
Personal services[a]	6.2	8.1	2.2	0.9	55.0	—	1.3	19.1	0.3	13.1	100.0
Occupation not reported	0.6	—	—	—	29.0	—	—	—	48.4	22.6	100.0
TOTAL NUMBER	100.0	242,352	52,164	9,882	361,584	1,944	13,932	63,180	10,530	75,978	831,596
PERCENTAGE		29.2	6.8	1.2	43.5	0.2	1.7	7.6	1.2	9.1	100.0

Source: Pennsylvania Department of Highways, Highway Planning Division. Data not adjusted to control information. Data for following occupations excluded from table: Housewife and housekeeper. Total 534,924; Miscellaneous: retired persons and students. 399,654; Other: Children and unemployed. 232,470.

[a] Includes personnel in Armed Forces.

TABLE III-5

PERCENTAGE DISTRIBUTION OF TRIPS BY PURPOSE, BY OCCUPATIONAL GROUP
CITY OF PHILADELPHIA, 1947

Occupational group	*Per cent of total trips*	*Purpose of trips*				
		Work and business	*Social and recreation*	*Shop-ping*	*Other*	*Total*
Professional and semiprofessional	6.9	68	19	6	7	100
Proprietor and managers	7.6	75	15	5	5	100
Clerks (store and office)	18.5	82	12	3	3	100
Skilled labor and foremen	14.7	77	14	3	6	100
Semiskilled and operatives	8.3	81	12	3	4	100
Unskilled labor	5.1	85	10	2	3	100
Personal service	4.2	79	12	6	3	100
Housewives	17.4	22	26	42	10	100
Other	17.3	31	29	12	28	100
All occupations	100.0	59	20	12	9	100

Source: U.S. Bureau Public Roads, Highway Transport Research Branch.

the site is the personal preference of the decision-makers in the situation. Consistency may also be noted for the upper extreme in median travel time, in that the longest trips by residents of both the city of Philadelphia and the entire SMA were made by persons in clerical occupations.

3.132. Modes Selected. Modes employed in home-to-work travel by gainfully employed persons in Philadelphia were measured in 1952-53 in connection with the State-wide motor-vehicle-use survey (Table III-4). The extremes of auto and transit use by occupation are the most notable elements of the data. It is not surprising that the greatest use of automobiles was by traveling agents and salesmen. Fairly comparable in their high rates of auto usage on work trips were the categories (in descending order of auto use) craftsmen and foremen, professional and semi-professional, and proprietors and managers. The occupations most dependent upon public transit were found to be clerks and nontraveling sales personnel, personal service workers, and semiskilled and unskilled workers. These occupational groups accounted for fully 52.6 per cent of the total workers represented in the sample data. At any rate, it is clear that over half of the employed persons were highly dependent upon transit as of 1952-53. It would seem that if transit companies were to conduct marketing surveys of their revenue customers as do most other enterprises,[4] then such information on the heavy users of transit would quite likely be found of more than theoretical interest.

3.133. Variations in Total Trips by Purpose. The differences shown in Table III-5 in their proportions of trips made for four major purposes by persons in the several occupational groups cannot be attributed to any special

[4] Examples of transit operators which have in fact made such surveys include the Cleveland Transit System and the London Transportation Executive.

characteristics of the work trips made by persons in these groups, but rather is more likely to be caused by the relative numbers of trips made for purposes other than work. For fairly comparable information on a national scale, but for the automobile mode only, the reader might refer to a marketing study by Alfred Politz Research, Inc., titled *The Automobile in the Daily Life of the American Population,* April, 1951, page 52.

3.14. Industry of Occupation

Little is known of the relationship of this characteristic of the labor force to patterns in the journey to work. Special tabulations for the present report had been prepared for work trips by place of origin, place of destination, and industry of occupation of worker. A sample of the data obtained was tested for Philadelphia (1947) against control information on located manufacturing employment for the year 1948. So little correspondence was found between amounts of manufacturing employment at the workplaces and the numbers of manufacturing work trip-ends in the respective traffic districts that further work on this aspect of the study was dropped.

This test is not considered to reflect negatively on any other part of the O-D survey in Philadelphia or elsewhere. Rather, it is believed that the apparent error lies in the subtle task of classifying industry of occupation in the original residential interviews, and that the most likely contributary cause could have been in failure of communication between interviewer and interviewee on what was being sought.

3.15. Ownership and Registration of Vehicles

The levels of ownership or registration of autos have been found to be closely associated with variation in daily trip generation when the trip data are reduced to a per capita or per dwelling unit basis. In particular, this point is stressed by the findings of studies in Detroit (1953) and in Washington, D.C. (1948 and 1955). As Table III-6

TABLE III-6

TRIP GENERATION PER DWELLING UNIT AS ASSOCIATED WITH FOUR VARIABLES[a]

WASHINGTON, D.C., 1948, AND DETROIT, 1953

	Correlation coefficient		*Standard error of estimate*	
Variable	*Washington 1948*	*Detroit 1953*	*Washington 1948*	*Detroit 1953*
Auto ownership	0.827	0.889	0.89	0.88
Net residential density	−0.718	−0.748	1.10	1.27
Income index	0.655	0.725	1.20	1.32
Mile-distance from CBD center	0.575	0.812	1.30	1.12

Sources: Washington, D.C., W. L. Mertz and L. B. Hamner, "A Study of Factors Related to Urban Travel," *Public Roads* (April, 1957), pp. 170-174; Detroit, *Detroit Metropolitan Area Traffic Study*, Vol. I. Data Summary and Interpretation. July, 1955. Table 39.

[a] Trips are those of all purposes. Simple correlation measures were computed for each variable in its association with trip generation.

shows, auto ownership has been found to be more closely associated with trip generation for all trip purposes than any of the other three major factors, including that of income level. How general this finding may be and how far into the future it may apply is not known.

3.16. Income Level

Section 3.1 indicates why the income level of families may be expected to be an important variable in the study of urban travel. In explaining variations in the generation of trips for all purposes from residences, income level has been found to be third in apparent importance in one city, and fourth in another, as may be seen from the correlation coefficients of Table III-6. The degrees of association found in the two studies were not very different for the income index; the difference in ranking of the variables appears because of a considerable difference between the two cities in importance of another variable, that of the distance of the traffic zone of the residence from the city center. The significance of the information in this table lies not so much in the difference in rank order of the third and fourth variables as in the almost surprising degree of comparability in the two independently conducted studies.

The Washington study indicated additionally that a correlation of 0.786 existed between income per household and automobile ownership. Thus, while the vehicle-ownership data appear more directly usable for traffic prediction, the auto-ownership and family-income index appear in some degree as measures of the same thing.

We have noticed that income has some relationship to the length of work trips (see Table II-9). The numbers of work-trip origins per dwelling unit throughout one city were also found to vary consistently with family income levels. For individual subareas within Philadelphia (1947), deviations from the city-wide average of 1.25 work-trip origins per dwelling unit were plotted against the respective deviations from the city-wide median family income of $2,869 (1949).[5] A consistent straight-line trend was found. To test whether or not the effects noted were functions of family size, the deviations in work trip-ends were plotted against figures on deviations from the city-wide average of 3.36 persons per dwelling unit (1947), for the respective subareas. In the latter analysis, no significant

[5] The number of work-trip origins averaged 1.32 per dwelling unit throughout the Philadelphia-Camden internal area in 1947. When internal trips classified as "purpose work to purpose work" are subtracted from the total, the ratio becomes 1.13 work trip-ends per dwelling unit.

relationship could be found. It was inferred that secondary employment contributed measurably to the relatively high median family income levels found in the northern neighborhoods of the city, since it is apparently secondary employment which swells the average number of work-trip origins in these districts.

The inference above concerning evidence of secondary employment is probably valid. However, the effect noted could be exaggerated relative to other areas of the city by two causes: (1) better interviewer-interviewee rapport and more accurate reporting of trips in areas of higher income, and (2) greater dependence of secondary employment in lower-income areas on walking to work. Perhaps both reasons were applicable, causing some exaggeration of the differentials, but even considered additively, the two reasons could not have accounted for all the observed effects.

3.2. CONSIDERING PLACE OF ORIGIN OF WORK TRIPS

In order to develop prediction factors for the land-traffic relationship, it is desirable to relate trip-ends to the land areas in which they are located. In studying the journey to work, it is desirable to consider the home-to-work direction, i.e., to consider the residence trip origin and the workplace trip destination. Reasons for this selection include the following: (1) The proportion of work-destined trips made directly exceeds the proportion of direct work-to-home trips. (2) Trips made to home from workplaces would tend to be confused with trips to home from nonworkplace origins.

Factors of interest for origin areas include: the location of the residential area within the urban area, its land-use composition, its density, and the character of the city or area of residence.

3.21. Factors in Total Trip Generation

By this time it should not be a startling concept that the number of total trips made in a day to or from a specific zone within or near a city has been found to be a function of such physical descriptive factors as the distance of that zone from some bench mark at the center of urban activities, and the intensity of use of exploitable space in the zone. Some indicative statistical measures of relationships in trip generation are presented in Table III-6. The matter may be pursued in considerably greater depth by studying publications of the Philadelphia, Detroit, and Chicago metropolitan area traffic studies and others which have become available in recent years.[6]

3.22. Work-Trip Lengths from Residential Origins: Intracity Aspects

A family of distribution curves of time-length of work trips by all modes, including walking, from a number of residential sections in one urban area has been assembled

[6] Among those which treat the subject in detail are:

American Institute of Planners. *Journal.* "Land Use and Traffic Models: A Progress Report." Special Editor, Alan M. Voorhees (May, 1959).

Detroit Metropolitan Area Traffic Study:

1. John R. Hamburg, *The Relationship of Distance, Residential Density, Relative Income, and Car Ownership to Average Trip Frequency as Reported at Households in the Detroit Area* (About 1955).

2. ———, *Land Use and Traffic Generation in the Detroit Metropolitan Area* (About 1955).

R. E. Schmidt and M. E. Campbell, *Highway Traffic Estimation* (Saugatuck, Conn.: Eno Foundation for Highway Traffic Control, 1956). See especially Chapter 2.

Highway Research Board, *Travel Characteristics in Urban Areas. Bulletin 203* (Washington, D.C., 1958). See also *Bulletin 224,* 1959, article by H. S. Lapin.

W. L. Mertz and L. B. Hamner, *op. cit.;* F. B. Curran and J. T. Stegmaier, "Travel Patterns in 50 Cities," *Public Roads* (December, 1958), pp. 105-121.

F. Houston Wynn, *Traffic Engineering Characteristics.* Vol. 1A. (Berkeley: University of California, Institute of Transportation and Traffic Engineering, 1955).

for study. The urban area of reference was Philadelphia and trip data are from the survey of the National Housing Inventory of 1956. All work trips by employed heads of households from residential origins are utilized in the data assembly, including trips made to destinations outside the city.

The two extremes, forming the envelope of the curve, and a curve which falls between, are shown in Figure III; basic data of the curves are given in Table III-7. Median travel times ranged from 23 minutes for the central business district, an area which contains some residences, to 40 minutes for the far northeast.

Utilizing the median lengths of work trips from residence, a three-variable family of curves has been prepared to illustrate a way of analyzing travel patterns in relation to city structure. The suggested approach is shown in Figure V. The travel-time information for the 12 areas (Figure IV) has been assembled with two other structural factors: net residential population density and the mile-distance of the area's centroid from the arbitrarily chosen central datum point of the city.

The contour method of assembling these associated variables gives promise of usefulness in the study of trip generation, since it permits simultaneous inspection of relationships among the variable of particular interest (in this case, the time-distance of work trips) and the two associated variables. Thus, what is suggested here is a means for visualizing coherent interrelationships which may be found in the attempt to relate travel behavior with aspects of urban structure.

3.23. Work Trips from an Urban Residential Area

Examination of travel patterns from a particular residential area within a city may be of interest for several reasons: (1) numbers of work trips may be compared

Figure III

THE DISTRIBUTION OF TIME-LENGTHS OF WORK TRIPS FROM RESIDENCES. PHILADELPHIA, THREE PLANNING ANALYSIS SECTIONS — 1956

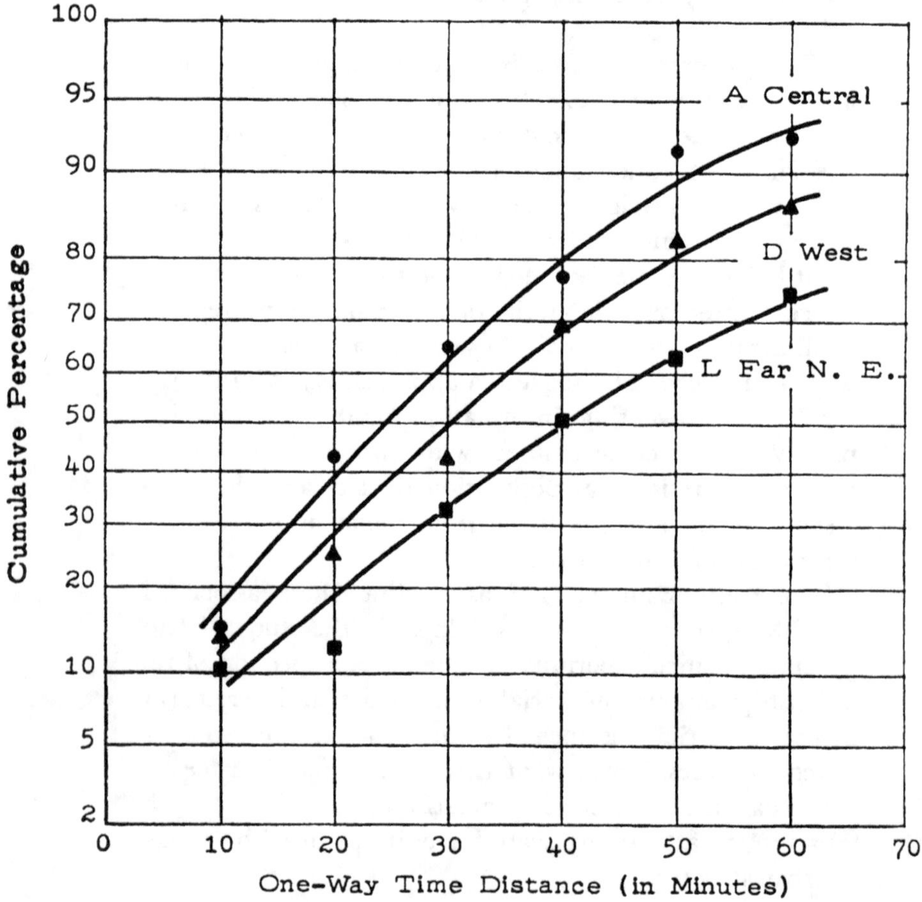

TABLE III-7

Percentage Distribution and Median Time-Length of Work Trips from Residence, by Employed Heads of Households

Philadelphia, 1956

Time length (in minutes)	*Central* A	*South Phila-delphia* B	*South-west* C	*West Phila-delphia* D	*Lower North* E	*Upper North* F	*Kensing-ton* G	*Rox-borough-Mana-yunk* H	*German-town-Chestnut Hill* I	*Olney-Oak Lane* J	*Near North-east* K	*Far North-east* L	*City*
0- 9	14.6	15.8	15.2	13.5	14.2	14.7	17.5	10.1	14.0	12.9	10.3	10.2	13.7
10-19	28.1	19.8	17.4	11.4	18.2	23.3	19.3	26.0	11.3	19.4	22.3	2.0	18.6
20-29	21.9	22.9	8.1	17.4	11.3	23.0	17.9	13.3	12.0	12.1	15.9	20.9	16.5
30-39	12.7	16.9	24.4	26.8	21.9	16.0	17.3	22.8	28.7	21.9	20.0	17.7	20.9
40-49	14.6	9.7	17.2	13.0	13.0	12.8	13.9	4.7	18.4	21.4	9.4	12.7	13.0
50-59	0.9	4.1	7.0	4.1	3.1	3.7	4.0	3.2	0	4.2	11.4	10.8	5.0
60+	7.2	10.8	10.7	13.8	18.3	6.5	10.1	19.9	15.6	8.1	10.7	25.7	12.3
TOTAL	100.0	100.0	100.0	100.0	100.0	100.0	100.0	100.0	100.0	100.0	100.0	100.0	100.0
TOTAL TRIPS	16,831	45,341	19,572	52,170	50,637	32,921	22,178	7,442	17,238	27,218	55,074	8,399	383,079
MEDIAN MINUTES	23.3	26.3	33.8	32.9	32.9	25.2	27.4	30.3	34.4	32.6	30.8	40.0	30.6

Source of basic data: U.S. Bureau of the Census, *National Housing Inventory*, Philadelphia Supplement, December, 1956.

Figure IV

PLANNING ANALYSIS SECTIONS
PHILADELPHIA

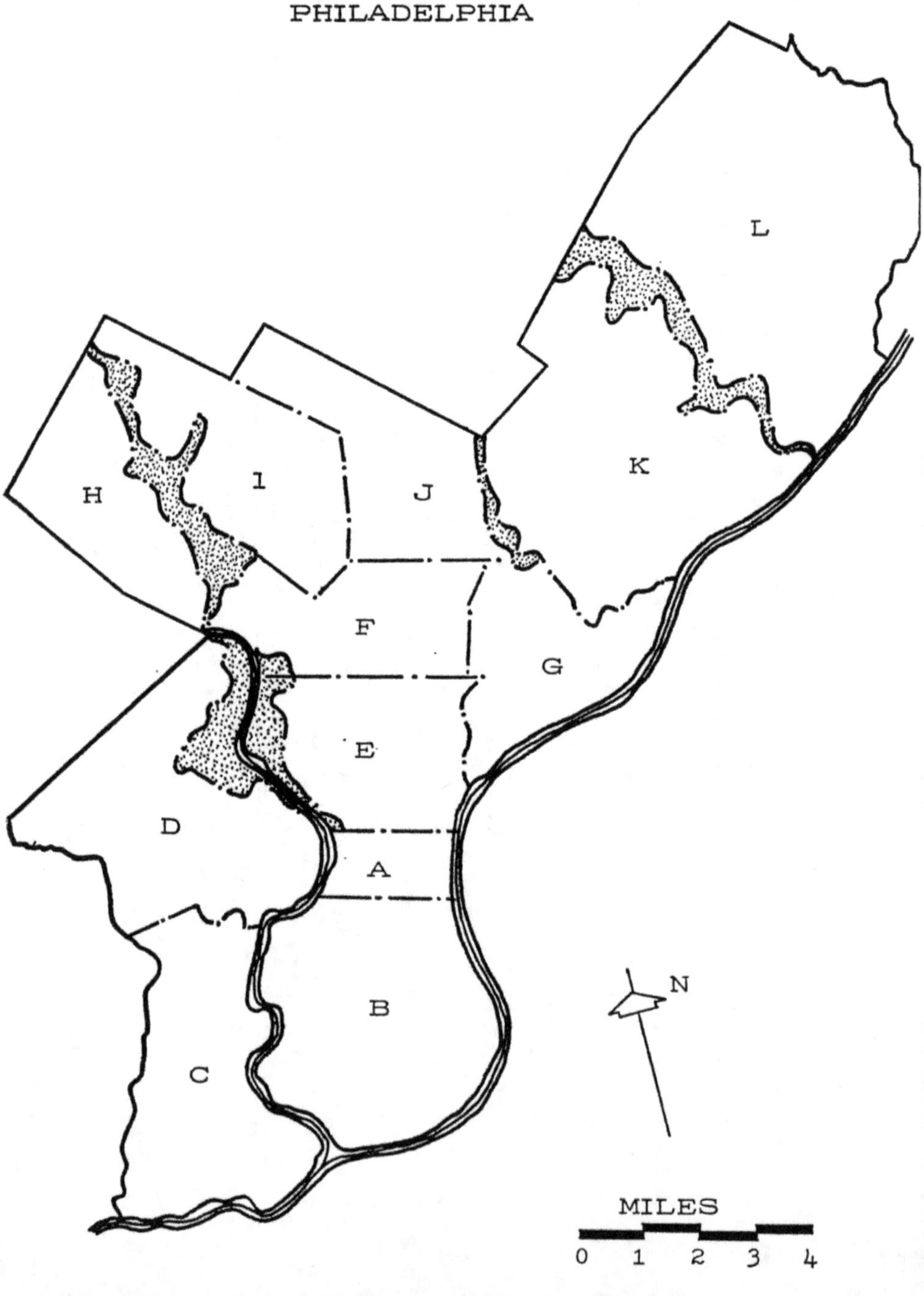

Figure V

SUGGESTED TIME-DISTANCE CONTOUR MAP. MEDIAN TIME-LENGTH OF WORK TRIP AS A FUNCTION OF RESIDENTIAL AREA CHARACTERISTICS PHILADELPHIA - 1956

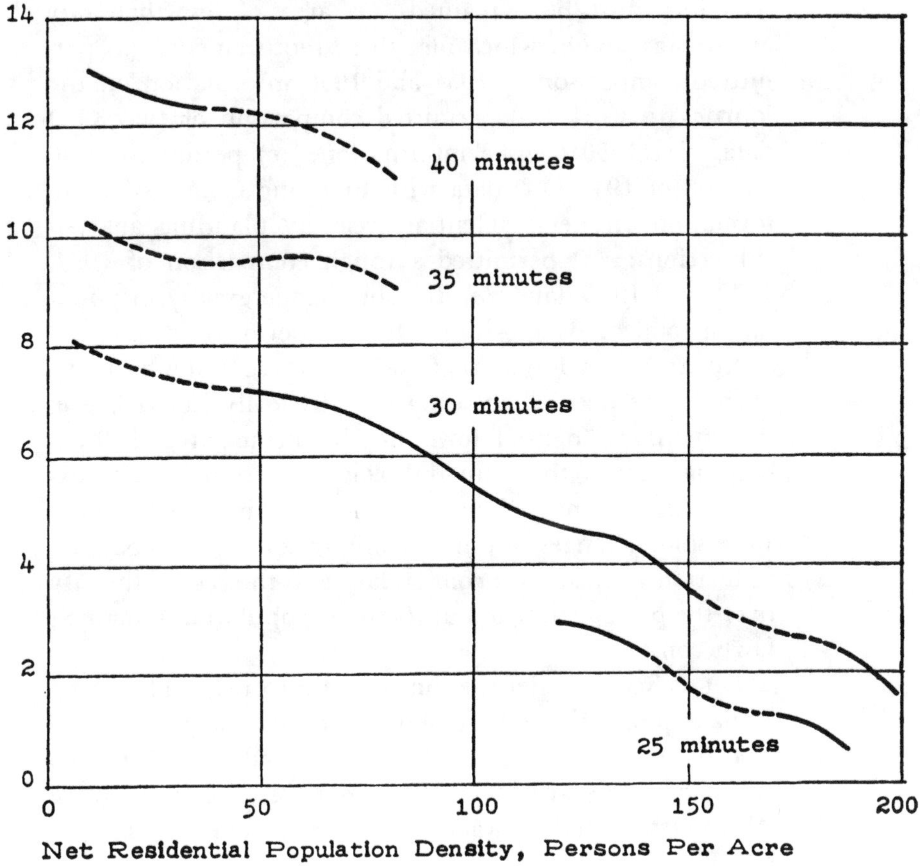

with total population and numbers of workers; (2) inference may be made concerning the data on trips made by walking, which were not gathered in the 1947 O-D survey; and (3) time trends in available data may be examined.

Since only one area is to be studied, rationale for its selection must be explained. An area of less than city extent was sought which met these requirements: (*a*) permitted comparison of 1934 and 1956 information on the journey to work; (*b*) permitted comparison of 1947 O-D data with 1950 Census information; (*c*) permitted comparison of 1947 O-D data with Planning Commission information on net residential areas by planning analysis subsections; (*d*) permitted a rough comparison of 1934, 1947, and 1956 data; (*e*) did not change greatly in population total or composition over the period studied.

No area was found that met all these criteria. However, it was decided to study the physically isolated, predominantly residential, lower-middle-income area of Philadelphia west of the Schuylkill River, Section D. The example chosen met all but criterion *e* fairly well. There were some changes in ethnic composition and a moderate reduction in median income relative to the rest of the city over the period studied, but the total population remained fairly constant.

Let us first consider the time-trend information for West Philadelphia. Table III-8 shows the percentage distribution of time-lengths of work trips of heads of households (primary wage-earners in 1934) in trips from residences. All modes, including walking, were considered in the two sets of information.

The cumulative distributions of these trips by time-length reflect the characteristics over the 22-year interval noted for the city as a whole. The curve for the 1956 distribution, plotted on a cumulative percentage basis on probability coordinate paper, decreased in slope from that

TABLE III-8

TRIPS TO WORK FROM RESIDENCES IN WEST PHILADELPHIA 1934 AND 1956

Time-length (in minutes)	Per cent 1934	Per cent 1956
0- 9	4.2	14.0
10-19	16.5	13.0
20-29	21.8	14.8
30-39	30.7	26.2
40-49	14.2	14.2
50-59	3.7	4.9
60+	8.9	12.9
TOTAL	100.0	100.0

plotted for the 1934 data. Thus, as found for the city generally, there were more very short trips and more very long trips in 1956 than in 1934. However, the median lengths did not change very much, being 32.5 minutes in 1934 and 33.1 minutes in 1956. The amount of change was insignificant in terms of a sampling survey of this type.

It will now be of interest to consider the relationship between O-D work-trip information from one part of the area, Subsection D-3, and comparable Census information.[7] In 1950, 19,120 paid workers resided in the 14,369 occupied dwelling units of the subsection. The average number of workers per dwelling unit was 1.33 in the Census year, and the average number of reported vehicular work trips

[7] Population centroid of D-3 is 3.4 miles west of the central datum point of Philadelphia (City Hall). Gross area is 0.96 square miles and net residential area (1950) 0.52 square miles. 1950 net residential density was 126 persons per acre. City average net density in 1950 was 100 and for West Philadelphia the net residential density was 111.

from origin in the 1947 internal survey was 1.28 per occupied dwelling unit.

Of the 19,120 paid workers, it was estimated that 18,688 would have to travel some distance between their homes and workplaces. This estimate was based on the assumption that 20 per cent of the self-employed earned their livelihoods at the same premises as their residences. Of those who had trips to make, about 6 per cent would have been sick, on vacation, or otherwise absent on the day of the O-D survey, and perhaps 9 per cent might have walked to work. The total number of vehicular work trips recorded as originating in D-3 on an average weekday (June to November 1947) was 18,334. The close agreement of this number with the estimated 18,688 is somewhat coincidental. What it seems to indicate is that the number of work trips in one day other than the "first" trips to work was about equal to the sum of workers who walked plus those who were absent. The average proportion of "from work-purpose to work-purpose" trips throughout the internal area averaged 16 per cent of total work trips. If this were the proportion of nonfirst work trips originating in D-3, it would indicate that 15,800 first vehicular trips were made; and this number of first vehicular trips would correspond with a total number of 18,500 traveling workers, on the assumption that 6 per cent were absentee on an average weekday and 9 per cent of those traveling to work walked to their destinations. The close agreement of the figures of 18,688 and 18,500 do not prove the several assumptions necessarily to be correct, only that they are consistent. But the finding of consistency in independently derived bodies of data, in the present state of the analytic knowledge in the field of urban transportation, is vital to any further development of the available information.

A percentage distribution of the mile-lengths of the vehicular work trips from origins in Subsection D-3 is shown in Table III-9 and Figure VI. The median distance of these trips as measured from the 1947 O-D survey data was 3.3 miles. Average speed on these trips is not known; however, the average speed of the automobile

TABLE III-9

PERCENTAGE DISTRIBUTION BY MILE-LENGTH OF VEHICULAR WORK TRIPS FROM ORIGINS IN SUBSECTION D-3

WEST PHILADELPHIA, 1947

Miles	*Per cent*
0-0.9	5.2
1-1.9	20.0
2-2.9	17.0
3-3.9	27.8
4-4.9	11.2
5-5.9	8.2
6 and more[a]	10.6
TOTAL	100.0

[a] Includes trips to external destinations.

work trips was estimated at about 11 mph, based upon off-peak travel times as measured in 1955 and adjusted roughly for time differences, by year, and by hourly peaking characteristics.

Since the median time-lengths of work trips from residences did not change significantly from 1934 to 1956, it seems logical to attempt to relate the median time-length of work trip (33 minutes) with the 1947 median mile-length (3.3 miles). Neglecting the omission of walking trips from the 1947 data, the apparent speed for the median length was 6 mph. This appears to be a low figure, particularly in view of the fact that inclusion of the 1947 walking trips would serve to reduce this speed. Not much

Figure VI

CUMULATIVE PERCENTAGE DISTRIBUTION BY MODE OF WORK TRIPS FROM A RESIDENTIAL ORIGIN AREA TO A DESTINATION IN AN INTERNAL AREA PHILADELPHIA, DISTRICT 062--1947

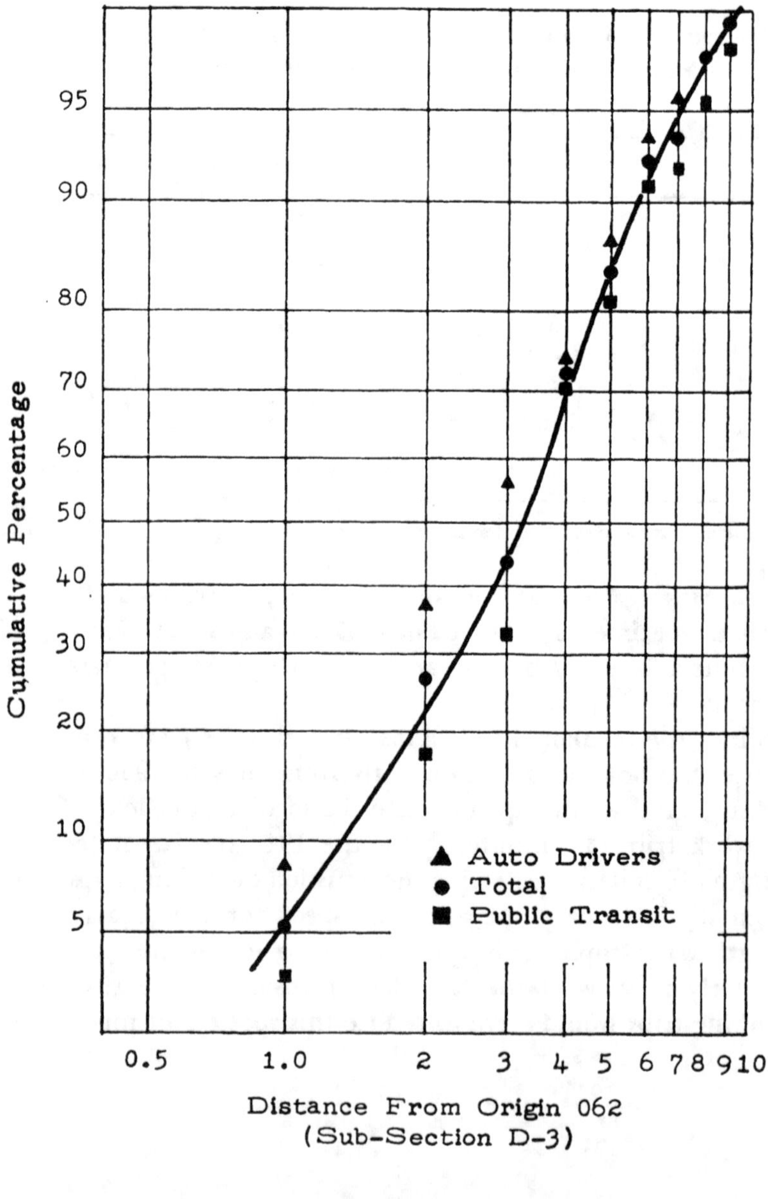

reliance is placed upon this derived speed. The speed would likely be somewhat above this figure for the trips of lengths longer than the median 3.3 miles.

The actual figures given for this example of study of work trips from a residential area of a city are not expected to be highly significant. Believed to be of interest are the criteria, the types of relatable information, and the limitations in the attempt to study work-trip dispersion from origins in a predominantly residential area.

3.3. SUMMARY

In this chapter we have considered work trips in terms of the persons who make the trips and of their places in the labor force, and we have noticed some relationships among work-trip patterns and the characteristics of the places where the trips begin. Mainly, factors of association rather than of direct cause and effect have been observed. That the relationships described are not cause and effect may be inferred from the fairly large number of items considered as associated variables with the few characteristics of the trips which are of interest.

Evidence of sufficiently consistent patterns to enable prediction of work-trip characteristics en masse appears in the descriptions given. However, the measures presented here are essentially superficial. We have glimpsed the ways in which work-trip characteristics vary, but have not really learned why. At the present stage of the art of prediction, it would appear that we could assemble building blocks in a way sufficiently similar to that of the original material studied to enable prediction. But very great care has to be taken in predictive efforts to insure that the superstructure of the forecast is founded upon a rational use of the original data of the base period. A rational use of the input data for past years should proceed from understanding of the subject phenomena, though it is true that for very short periods of forecast a fairly superficial approach to the problems may yield results of sufficient quality to be useful.

IV

Confluence at Workplace Destinations

CHARACTERISTICS OF WORK TRIPS HAVE BEEN CONSIDERED IN terms of persons making the trips and of the places from which the trips originate. In this chapter it remains to consider work trips and their patterns in relation to their destinations, and to consider ways for incorporating observed detail into general schemes.

This chapter describes several gross characteristics of communities and of workplaces relevant to study of the journey to work: urban functions and trip generation, patterns of travel from origins to workplaces, and means for synthesizing information on trip dispersion from places of origin (residences in most cases) and confluence at workplaces into an analytical methodology.

4.1. TRAVEL PATTERNS AND CHARACTERISTICS OF CITIES

Patterns of travel in a city are affected by and in turn help to form the city's structural characteristics. It has been mentioned that the individual decisions on the locations of homes and workplaces in the aggregate represent strong city-forming forces. But at any point in time, the existing physical structure of the city conditions the locations of trip-ends, the lengths of trips, the modes, and other elements of trips made by individuals. Similarly, the type of traffic generated is presently directly affected by the mix of land uses in a particular zone but, in the long run, patterns of traffic will affect types and intensity of uses in the zone.

The emphasis, here and elsewhere in this report, upon one aspect of the relationship—on the physical form of cities as it affects urban travel—should not cause the reader to forget the concept of the long-run interaction, i.e., the long-term effect on city form of patterns of urban travel.

Those things which contribute to shaping the daytime population pattern, and hence to forming patterns of urban traffic, include the economic character of the community and its place in the regional economy; characteristics of the area's circulation system; and the spatial distribution of commercial and industrial activities. In this regard, particularly in reference to the latter types of economic activities, we are led into measuring certain types of distribution densities: e.g., employment densities in industrial areas, and trips terminating per square foot of nonresidential floor space in the central business district.

Other factors affecting the distribution of daytime bases of operations, but not so subject to effective measurement, are those which may be reflected in the attitudes of the local population, in their vitality, and in special regional characteristics of the urban area. More material factors having their effects include the natural features of the site, the topography, subgrade conditions, climate, and relative availabilities and prices of fuels for vehicle propulsion.

4.11. Character of the Community

Certain of the functions of cities register obviously in the distribution of activities. An area with greatest strength in finance, trade, and service is likely to have a highly developed, dense central business district. On the other hand, a primarily manufacturing city is likely to have a weak CBD, with outlying centers that develop intense concentrations of automobile traffic.

Attention has already been given in preceding chapters to differences in composition of industrial employment and to some observable variations in travel patterns for four urbanized areas (Sections 2.41 and 2.42. Note also Appendix C, Tables C-8 and C-9, and Section C.323). In that consideration, only a few gross descriptive measures concerning urban travel were noted in relation to industrial composition. The actual effects of economic character on travel patterns in metropolitan areas are believed likely to be far more important than has yet been defined or roughly measured.

It is in terms of economic functions, as measured by proportions of persons in certain key occupations, that the Federal Aviation Agency (F.A.A.) has proposed generalizations about the character of communities. The purpose of such classification is to aid in analysis of the generation of intercity air travel, with an entire metropolitan area regarded in the way we view a local traffic district. The F.A.A. kind of classification does not have any immediate applications for explaining differences in travel patterns among cities, but it is possible that intercity comparison by means of key variables (not necessarily those selected by F.A.A.) may eventually prove useful in this regard.

So far, the discussion has referred only to the economic character of communities. The reference proceeds from the greater ease of relating quantitative economic data to noticeable effects in traffic patterns, as well as from time and space limitations on the scope of the present report. Certainly other characteristics, cultural, demographic, etc., must contribute to the differences among cities in the kinds of urban travel that occur. The point concerning classification of cities for study of characteristics was well stated in *Economic Geography.*

Cities serve manifold functions in the economy and culture of a people. All cities have some functions in common; all cities have some functions that are peculiar to their site and situation, to the people whom they serve; and all cities have some functions peculiar to their development and their history; hence cities may be classified more effectively on the basis of their functions as criteria than perhaps according to any other attribute. Size, ethnic composition or occupational activity of their people, topographic location, and accessibility, have formed bases for the grouping of cities for particular purposes.[1]

4.12. Daytime Population Distribution

Workplaces and other commercial establishments comprise daytime residences for more than a third of a city's population. Thus the periods of travel between homes and workplaces—the peak periods—represent the transitional intervals of time between the distributions of the population at their Census-indicated or nighttime residences and at their daytime residences.

It is reasonable to suppose that a greater measure of understanding of urban structure and of urban travel would result from detailed study of the relationships between the distributions of nighttime and daytime population densities. One important study has been made in this regard.[2] Carefully controlled statistical examinations were conducted for the urbanized areas of Erie, Pa.; Grand Rapids, Michigan; Minneapolis-St. Paul; and Philadelphia-Camden. Basic data used in this effort were from

[1] April, 1945.

[2] *Population Distribution Spatial and Temporal.* A study of daytime and nighttime differentials in the proportional distribution of the total population of selected urban areas. Industrial Areas Study, Institute for Research in Social Science (University of North Carolina, September, 1952).

origin-destination surveys conducted by the respective state highway departments according to the form established and recommended by the U.S. Bureau of Public Roads.

It was found that generalizations applicable to patterns in all the cities studied could not be made with a sufficient degree of statistical reliability. "No norms of population variation for different cities or for different subareas of cities have been clearly identified, and without these the investigation of patterns of causal relationships between population distribution and fluctuation and urban characteristics is exceedingly difficult." However, as the authors point out, for an individual urbanized area, the study does serve as a first example of a means for locating population by functional land-use categories, by time of day.

The authors of the study suggested that the rate of growth of a city's population might be a more significant key to understanding population distribution than the mere numbers of persons present at a given time. The suggestion was also made that greater understanding of patterns in population distribution might be gained through study of the relation between "travel time and residential, industrial, and commercial location, or the analysis of residential location of important economic groups in relation to workplace."

This was a valuable pioneer study of an important problem.[3] It was rigorous in statistical methodology. The fact, however, that the cities or urban areas considered were of widely varying population sizes leads to a question as to whether or not some means of correcting for the scale factor might not be introduced, in the effort to find a greater degree of comparability in the raw data.

The study of daytime population described above was

[3] Another excellent early work on this subject was by Gerald W. Breese, *Daytime Population of the Central Business District of Chicago* (Chicago: University of Chicago Press, 1949).

based upon located destination trip-ends. Another approach to locating daytime population is illustrated by an effort made in Philadelphia. A complete study was made by the local civil defense office in cooperation with technicians of the U.S. Bureau of the Census, to locate the daytime population in Philadelphia. A portion of the data compiled and calculated are shown in Table IV-1. Shown

TABLE IV-1

DIFFERENTIALS BETWEEN RESIDENT AND ESTIMATED DAYTIME POPULATION FOR A SMALL NUMBER OF CENSUS TRACTS PHILADELPHIA, 1952

Census tract	*Resident population*	*Daytime population*	*Net daily migration*	*Per cent change*	*Public school accumulations (net)*
City	2,071,605	2,466,284	394,679	19.05	—
1A	589	3,181	2,592	440.1	—
1B	15,014	8,647	−6,367	−42.4	822
1C	16,890	11,844	−5,046	−29.9	1,950
2A	9,323	7,670	−1,653	−17.7	571
2B	9,413	9,141	−272	−2.9	606
2C	1,616	11,796	10,180	630.0	—
3A	10,012	8,011	−2,001	−20.0	285
3B	2,256	2,431	175	7.8	1,571
4A	8,075	13,264	5,189	64.3	1,003
4B	1,979	3,195	1,216	61.5	487
5A	6,982	37,253	3,027	433.6	732
6A	1,031	58,143	57,112	5539.5	—
7A	5,913	9,145	3,232	54.7	—
7B	5,469	6,597	1,128	20.6	—
7C	4,171	2,174	−1,997	−47.9	—

Sources: U.S. Census of Population, 1950; U.S. Census, special compilation, June 30, 1952.

are the estimated daytime population, the nighttime or residential population, and the calculated daily migration, given in percentage change between daytime and nighttime population. It is important to note whether the increase in daytime population is caused by the presence of a school.

The information of which Table IV-1 represents a part was compiled from data of the Census of 1950 and from a special census prepared for the Federal Defense Agency by the U.S. Bureau of the Census.[4] Primary factors in the compilation include the numbers of resident workers who left home to pursue their occupations and the numbers of school-age children.

The disadvantage in the Philadelphia method is that validity of the results cannot be determined, since no thread of connection exists between nighttime and daytime figures to indicate who went where. One reason why it would be desirable to include journey-to-work information in the regular decennial census would be to obtain comparable tabulations on locations of daytime versus nighttime bases of operation.

4.13. The Population Profile of Cities

The population profile is a valuable concept in describing the spatial distribution of a city's population outward from the geographic center. It refers to the varying pattern of settlement with increasing distance from the center, and is designed to allow comparisons among cities. The profile shows the proportion of the city's population in successive mile-rings of distance from the center, and is prepared by plotting percentage of population for each mile-ring on the vertical scale against a linear horizontal scale representing mile-rings of distance from the center. Profiles of this type, in terms of total resident population, have been drawn by F. Houston Wynn for a number of American cities.[5] Density is not a usable term for this profiling concept as the mile rings about the center are not of equal area.

[4] Survey was dated June 30, 1952.

[5] F. Houston Wynn, *op. cit.*, Figure 10.

This profile, however, can be drawn as well for the distribution of employment and school, or daytime, population. Superposition of daytime and nighttime profiles roughly reflects the net distance effects of the daily person trips from residence. The horizontal distance between modal points of the two curves has meaning as an index of aggregate distance between intensively developed areas of residence and of economic activity. Comparison of the double profiles for a number of cities in similar population classes might be helpful in gaining understanding of the general scope of circulation problems confronting individual cities.

One disadvantage of this measure, of course, is in the unequal areas of the rings: the greater the distance from the center, the greater is the area covered by the individual mile-rings. An alternative method would be to compare profiles for various cities, using equal-area distance rings. Such distributions could be called density profiles. For an individual city, the profile obtained by this method would be less understandable to the casual observer than analysis by even-interval mile-rings, but the equal-area rings might well have more meaning for intercity comparison, particularly if net areas were used.

4.2. CHARACTERISTICS OF WORKPLACES

The characteristics of workplace areas of interest here include the volume and the density of employment, the location of the workplace with respect to the business center of the urban area and to the surrounding urbanized area, and the types of economic or industrial activity. The type of economic activity, of course, relates to the characteristics of the labor force, and to matters discussed in the preceding chapter.

4.21. Volume and Density of Employment

The numbers of employees and the density, as measured in employees per unit area of a workplace, are considered as related facets of the scale factor. This consideration seems desirable since the relative importance of number and density as trip-generating correlates seems to vary from one site to another (possibly because of the lack of uniformity in the use of the term density in reference to employment).

Number of employees is considered first. The closest approximation to a controlled study of the effects of numbers of employees is that made of travel patterns to work of workers at six plants of a large automotive manufacturing concern in Flint, Michigan.[6] Total employment was about 66,000 and the plants varied in number of employees from about 3,000 to 25,000. Schnore observed that the larger plants drew workers from the greater distances. When, however, the numbers of workers traveling to each plant from each circular distance zone around it are considered as proportions of total resident population, the differences in drawing power among five of the six plants became much less significant. (The case of the sixth plant is discussed below, in Section 4.23.) From this it appears that differences in the sheer volume of employment at the five plants was not as significant a factor as appeared on first inspection. Summary inference from the scale-adjusted data is that the probability of workers' traveling from greater distances was about the same for the five plants, but that actual occurrences of longer trips were greater in absolute numbers for the larger plants. Volume of employ-

[6] Leo F. Schnore, *The Separation of Home and Work in Flint, Michigan.* (Ann Arbor: Institute for Human Adjustment, University of Michigan, 1954).

ment is thus a characteristic to be observed but, wherever possible, eliminated from consideration by arithmetic or statistical means.

Two concepts of density may be considered. The first concerns the number of employees per unit area of the destination site, and the second concerns the number of trip-ends at the site. The two concepts are related, of course, but the distinction between them should be observed. Employment density is a general descriptive measure of the workplace site, whereas the trip-end density varies with the system of trips under consideration. That density of employment and density of work trip-ends (considering the system of work trips to a particular destination) are closely related measures is fairly obvious. It is this relationship, in fact, which was the basis for the suggested method of analyzing daytime population density made by the Institute for Research in Social Science of the University of North Carolina (see Section 4.12). This distinction in concepts is emphasized here, however, to indicate the possibilities of analyzing destination trip-ends for systems of trips made for other purposes than going to work.

It is suggested here that density of employment may be a relevant factor in determining the distribution of origins of work trips around a workplace destination. To what extent such density may differ in its effect from that of location of workplace or distance from a central benchmark is not known. The two factors must certainly be related. At the present stage of lack of knowledge of its effects, two kinds of people-per-acre measures for employment are suggested for further study (see *Glossary*).

The importance of employment density is indirectly implied in work reported by the late Sam Osofsky, formerly Highway Economist of the California State Division of

Highways and a great (and insufficiently appreciated) innovator in the field of highway traffic analysis.[7] Osofsky's trip prediction model, for trips made for all purposes in the Vallejo Area, utilized the following characteristics of zones studied: numbers of employees, commercial acres, industrial acres, straight-line distance between zones, and the product of multiplication of volume of population and owned vehicles in the zones. In this assembly of data, the figure for number of employees is raised to the second power.

Implicit in the first three of the characteristics employed in the Vallejo trip prediction model is the concept of employment density—numbers of employees and acreage of the workplace site—even though the data may not have been cranked into the machine for the purpose of employing the density concept. He found the trip model based upon these variables to be statistically sound for description of present travel and to be commutative over time (discrepancies unrelated to quality of forecasts were kept reasonably low in models prepared for a forecast year).

One study, particularly directed to measuring the generation of truck trips, noted significant relationships between truck trip-ends and the density of manufacturing workers at workplace sites.[8] Not a great deal more may be said specifically about employment density as it may be associated with patterns of travel, but the evidence is strong that this type of measurement presents an avenue for future useful research in urban transportation.[9]

[7] Osofsky, *op. cit.*

[8] Hans Blumenfeld's work for the Philadelphia City Planning Commission, *circa* 1954. See footnote c, Table IV-3.

[9] Students pursuing this direction should notice the great difference in space demands by industry, by size of city, and by character of city. Reduction of mass data to standardized units is a first step. A good reference for early consultation in such work is Harland Bartholomew, *Land Uses in American Cities* (Cambridge: Harvard University Press, 1955).

4.22. Location of Workplace

Location of workplace, in a few instances, has been found to be of direct relevance to patterns of interzonal travel. A study in Detroit reported that: ". . . it is possible to predict the true interchange between two zones within extremely close limits by knowing only four things: (*a*) the total trips originating in each zone, (*b*) the total trips made in the area, (*c*) the airline distance separating the zone centers and (*d*) the distance of each zone center from the Central Business District."[10] Location of workplace has also been found relevant to choice of modes for travel to work in Philadelphia. The latter indication is discussed in the following section.

Patterns of trip origins about workplace destinations change with the location of such destinations, largely because of the varying residential density patterns by milering around each focus throughout a metropolitan area. This type of effect of location may be inferred in graphical analyses.

Carroll has distinguished between the residential distribution of persons employed in central business districts and those who are employed in workplaces in intermediate or peripheral locations.[11] He states that the residential distribution of the former approximates that of the population of the entire urban area, and he points out that "residences of persons employed in off-center (intermediate or peripheral) workplaces are concentrated most heavily in the immediate vicinity of the place of work." This finding is substantiated in the DMATS study for Detroit (1953).[12] Travel to four noncentral factories as well as to the CBD was studied. It was found that the four plants "draw more

[10] Detroit Metropolitan Area Traffic Study, *op. cit.*, p. 100.

[11] J. Douglas Carroll, Jr., "The Relation of Homes to Work Places and the Spatial Pattern of Cities," *Social Forces*, XXX (March, 1952), 271-282.

[12] DMATS. *Data Summary and Interpretation*, Part I, pp. 95-97.

heavily from residential areas which are on the opposite side of the plant from the center of the city." Employment at the four plants did not differ much (first trips to work varied from 5,400 to 8,700), and density is not known. The scale of the CBD employment was much greater, first trips from internal origins to that locus being of the order of 80,000. The attraction pattern to the CBD was quite different from that of travel to the four Detroit plants. So far as the uniqueness of the CBD pattern is concerned then, location (and thus accessibility) did appear relevant but probably was not the only contributing factor, since scale, type of economic activity, and possibly employment density differed from similar measures for the manufacturing plants.

From the evidence given, no blanket statement may be made about the importance or direct relevance of the locations of workplaces for analysis of travel in all cities or metropolitan areas. It is possible that in some areas other related factors, such as employment density, could explain a greater proportion of the variation found in measurements of comparable characteristics in patterns of confluence at workplaces. Several avenues of research on the effect of location are indicated in the Detroit study mentioned above. Another is suggested by the work of Ernest Jurkat, in his studies of the relationships between residential population densities and distances from a central bench-mark in a given metropolitan area.[18] Such consideration might also be given to employment densities and to their rates of change over a period of decades.

In order to attempt further generalization about the

[18] See "Transportation Requirements—Population and Employment," *Technical Memorandum Number 3*, Philadelphia Urban Traffic and Transportation Board, 1956; also, Jurkat's article in the April, 1957 issue of *Traffic Quarterly*. On this subject, reference might also be made to "Transportation Analysis in the Philadelphia Metropolitan Region" by H. S. Lapin. *Papers and Proceedings of the Regional Science Association*. Vol. III, 1957, pp. 198-206.

effect of workplace location in the metropolitan area, it becomes desirable to establish some categories of relative location with respect to a central datum. The radial scale of distance may be measured outward from the center in mile- or time-distance, or along a spectrum of values of residential density. For the present purposes, these four categories are found useful: central, intermediate, peripheral, and outlying. Central refers to a location within or close to the central business district; intermediate, a location outside the CBD but within the urbanized area. Peripheral workplaces are those on or close to the margins of contiguous developed area, and outlying centers, those separated from the developed or urbanized area by intervening open or agricultural land or by major obstacles in the terrain.

The distance and direction of outlying places of employment relative to the CBD are both important. The trip distribution pattern is affected by the area from which the destination draws employment, since those persons who must cross the central business or other dense area of the city have an added friction of travel. Therefore, the distribution pattern on the side of the plant with higher density may be expected to be of a shorter mile-length. For outlying plants drawing workers from the open country beyond the city, mile-distance of trips may be expected to increase considerably beyond that of persons commuting through or around the city. In areas not highly built up, as in the case of destinations studied in Wisconsin and in up-state New York, there are found numerous instances of one-way work trips of 60 miles and more in length.[14]

Thus, in addition to location of destinations by mile-

[14] George P. Stevens, Jr., "Sample Study of Residential Distribution of Industrial Workers in an Urban Community." *Land Economics,* XXVIII (August, 1952), 278-283; Howard E. Conklin, "The Rural-Urban Economy of the Elmira-Corning Region." *The Journal of Land and Public Utility Economics,* XX (February, 1944).

or time-distance rings about the CBD or other central datum of an urban area, location of the destination by directional sector with respect to the center takes on significance. The sector location of destinations is of meaning to the extent it is desired to explain asymmetrical distributions of origins or asymmetrical patterns of work-to-home isochrones about a noncentral destination.[15]

4.221. Mode and Location of Workplace. In one study, a clear relationship was found between the mode selected for a trip to work and the distance of the workplace from the center of the metropolitan region. It was found in the Philadelphia-Camden internal area (1947) that for workplaces at three miles distance from the regional center in Philadelphia 40 per cent of person work-trips were made in autos. At seven miles distance the proportion was 60 per cent, and at 11 miles the proportion was 70 per cent. The data described, based upon about 800,000 trips made to 67 subsections, plotted in a distinct linear pattern on semi-log coordinates, with the distance expressed in logarithmic terms. There were only a few points of notable deviation observed in the study.

4.222. Car Occupancy and Location of Workplace. Car occupancy on work trips arriving at specific areas was studied in relation to distance of the workplaces from the city center.[16] In this assembly of data (for Philadelphia, 1947) no structured relationship could be observed. Car occupancy is discussed on a more general basis in Section 2.5.

[15] A sector analysis of residences about a central destination is given by Jacob Silver of the U.S. Bureau of Public Roads in the article, "Trends in Travel to the Central Business District by Residents of the Washington, D.C. Metropolitan Area, 1948 and 1955." *Public Roads,* XXX (April, 1959), 153-176.

[16] Data for this relationship were assembled by personnel of the Philadelphia Zoning Advisory Commission to test assumptions regarding off-street parking requirements in a proposed zoning ordinance. Special work-trip tabulations of the present study and instructions for their use were furnished by the author. Miss Marcia Rogers, now with the Philadelphia City Planning Commission, prepared the graphical analyses.

4.23. Types of Economic or Industrial Activity

Trip production or attraction in terms of suitable units of measurement has been found to be related to the uses of the site at which the trip begins or ends. This concept was well and fully stated by Mitchell and Rapkin.[17] Joseph T. Stegmaier, first in a doctoral thesis at Johns Hopkins University and later with other staff members of the Bureau of Public Roads, has also observed varying levels of trip generation for different types of land-occupying uses.[18]

This concept is to be treated here in terms of the central business district when considered as a focus of work trips and, in a more limited sense, the question is to be raised as to whether there may be differences in work-trip patterns by types of economic or industrial activity.

Considering the downtown area of Boston, Robert Davidson grouped the available information on nonresidential floor space into four major categories: retail and consumer service, office, manufacturing and utility, and wholesale and storage facilities.[19] Utilizing information on trips for all purposes, Davidson found significant differences in the number of trips attracted per unit area of these four generalized categories of economic activities. The ratios obtained applied in a single year, that of the Boston origin-destination survey. He found that accommodations containing retail and consumer services attract over twice as many persons per square foot as office accom-

[17] *Op. cit.* See Chapter VIII, "Relating Movement of Persons and Goods to Land Use: Selected Explorations," and especially the table on page 154 and the figure on page 176.

[18] Joseph T. Stegmaier, *A Study of the Traffic Generated and the Parking Demand Created by Buildings with Various Types of Space Usage* (Unpublished thesis, Johns Hopkins University, May, 1948).

[19] Robert G. Davidson, *The Travel of Persons to Downtown Boston* (Unpublished thesis in partial fulfillment of requirements for degree of M.C.P., Massachusetts Institute of Technology, Cambridge, 1954).

modations. Manufacturing facilities generate about the same number of trips per unit area as do offices, which, of course, attract more person-trips than do wholesale-storage facilities.

Considering gross land-use categories, the DMATS found clear differences among them in the proportions of work trips to all terminating trips (1953). Selected Detroit data are given in Table IV-2.

TABLE IV-2

PROPORTION OF WORK TRIPS TO ALL TERMINATING TRIPS, BY LAND USE[a] DETROIT, 1953

Land use	*Per cent*
Commercial	32.2
Other business	66.2
Light industry	64.3
Heavy industry	81.6
Institution	33.1

Source: DMATS, *Data Summary and Interpretation*, Part I, 1955, p. 81.
[a] Internal trips only. Destinations throughout internal area.

Data for the Philadelphia central business district were prepared to indicate the densities of work trips and of all trips from the internal area which terminated in the district.[20] Proportions of work trips to all trips terminating in the district were also calculated. However, the type of data available forced comparison of densities and propor-

[20] The Philadelphia central business district is generally defined as the district of over two square miles south of Vine Street, north of South Street, and between the Schuylkill and Delaware Rivers. The highly developed commercial core of this district is roughly coterminous with Traffic District 000, which is delineated by the following four streets: 7th, 18th, Arch, and Locust.

tions by ten traffic districts within the CBD, rather than by functional categories for all nonresidential floor space in the CBD, as in the Boston study. Data for the ten districts of the Philadelphia CBD are shown in Table IV-3.

The varying attractiveness of destinations in downtown Philadelphia by type of economic activity may be inferred from the data of Table IV-3 to the extent that a dominant economic character is ascribable to each of the ten traffic districts. A notable range in the attractiveness of nonresidential floor space in the CBD is indicated. District 009, for instance, at the time of the survey was an area characterized by wholesaling with stocks, warehousing, and some manufacturing. It had a relatively low rate of attraction in total trips (175 internal trip-ends per 100,000 square feet of nonresidential floor space, or only 60 per cent of the CBD average). On the other hand, the core area of the CBD, District 000, with most of the office and major retailing facilities, had a trip-attraction ratio of 379 trips per equivalent area, a figure 30 per cent greater than the average for the entire CBD. It may also be noted that 85 per cent of the trips ending in 009 were made to work, whereas only 50 per cent of those in 000 were for this purpose. This difference is accounted for, of course, by the character of core District 000 as a personal service and retailing area, as well as a center of office-type economic activities.[21]

Comparison may be made between the last column of Table IV-3 and the figures of Table IV-2. In Philadelphia District 001, an area of warehousing and manufacturing, 83 per cent of the terminating internal trips were made to work. This proportion may be seen to be very close to the Detroit proportion for heavy industrial areas. The core of the CBD, comprising what might be called "commercial"

[21] For an excellent analysis of relationships among land-using establishments in the metropolitan center, the reader is directed to John Rannells, *The Core of the City* (New York: Columbia University Press, 1956).

TABLE IV-3

DESTINATION TRIP-ENDS AS A FUNCTION OF NONRESIDENTIAL FLOOR SPACE[a]

TEN TRAFFIC DISTRICTS OF PHILADELPHIA CBD, 1947

	Trip-ends, all purposes			*Work trip-ends*		
District of destination	*Total*	*Net*[b]	*Per 100,000 square feet non-residential floor space (A)*	*Total*	*Per 100,000 square feet non-residential floor space (B)*	*Ratio B/A*
000	310,640	245,250	379	121,820	188	0.497
001	8,640	8,460	263	7,020	218	0.830
002	31,300	29,910	307	12,230	126	0.409
003	6,580	6,510	336	3,720	177	0.571
004	16,510	14,890	214	6,440	93	0.433
005	36,070	25,370	136	18,190	98	0.717
006	16,770	15,880	302	10,560	201	0.665
007	8,310	8,140	166	6,910	141	0.849
008	15,090	14,900	206	11,570	160	0.777
009	19,320	16,270	175	13,870	149	0.852
CBD[c]	469,230	385,580	292	212,330	161	0.551

Sources: Trip-end data from 1947 O-D survey, as published in *Person-trips to the Central City*, Philadelphia City Planning Commission, 1954. Floor space data compiled from raw data supplied by same agency. Raw data applicable to year 1948.

[a] Internal trips only.

[b] Destination trip-ends of two-purpose categories eliminated: serve-passenger, and change-travel-mode.

[c] Utilizing the 1947 O-D and 1948 floor space information, Hans Blumenfeld found an average of 34 truck destination trip-ends per day per 100,000 square feet of nonresidential floor space for the Philadelphia CBD. (From an unpublished study prepared for the Philadelphia City Planning Commission.) It may be interesting in the present analysis that the St. Louis Plan Commission has found a measurable relationship (1957) between numbers of commercial-vehicle trips at workplaces and an index of employee density, or number of manufacturing workers per gross acre of site. The distinction may seem too slight, but in order to maintain consistency with other studies, the term employee density is used here only in the sense of the previous sentence (in which, for instance, office workers at a manufacturing plant are excluded from the calculation of the density ratio), and employment density is used to refer to a ratio of total employees per unit of workplace site, regardless of industrial or occupational category.

and "other business" uses, had a proportion of work trips to total at the 50 per cent level, or about halfway between the percentages given for these two use groups in Detroit. These comparisons are obviously rough, but such evidences of similarities in urban structure lead to moderate optimism about the potential for synthetic forecast models of urban travel patterns.

Insofar as spatial patterns of work trips are concerned, the type of economic activity of the workplace is not to be considered so much a direct cause of variation as an envelope for variables associated with differences in travel characteristics. One aspect of this concept may be seen in a partially controlled study: controlled in the sense of holding constant some of the variables.[22] In Schnore's study for Flint, referred to in Section 4.21, all six plants studied are located in the same urbanized area, all engaged in manufacturing automotive parts or assemblies, and all are under one ownership.

In this case, the factor most strongly affecting trip characteristics was found to be the composition of the employed force; and the determinant of the proportion of men and women employees was the type of manufacturing activity of the plants. Of the six plants, three were located centrally in the city, and three were in peripheral locations. However, five of the plants had essentially similar patterns of attraction as measured in terms of the proportions of resident workers traveling to each plant from equivalent distance rings drawn about each plant. The attraction of the sixth plant as measured in this manner decayed with increasing distance at a much faster rate than was found for the others. Women made up 48 per cent of the employed force of the atypical plant, while the percentage of women employees of the other five plants was far lower. At this

[22] Leo F. Schnore, *op. cit.*

one plant, where nearly half the employed force were women, nearly 70 per cent of all employees walked to work, while the average proportion of walkers at the other five plants was about 25 per cent. In this example, in a metropolitan area of moderate population size (271,000 in 1950), the significant factor appeared to be the distinction among types of manufacturing (light or heavy goods) and the resulting effects on the relative proportion of female employees to the total. It would have been interesting to know what proportion of the women employees who walked or used transit were secondary workers, as compared with primary wage-earners in their families, the latter of whom apparently were the much heavier users of autos in the journey to work.

Another study of manufacturing plants in Detroit indicates the proportional distributions of time- and mile-distances between homes and workplaces for over 87,000 employees. A neat summary was prepared to indicate patterns of residence about seven major plants in that city by the Detroit Metropolitan Area Regional Planning Commission.[23] One additional important factor is indicated: age of the plant. This relates to the length of time the employees have had to adjust their residential locations for convenience of travel to work. The text makes no direct statement on the effect of plant location, but such information might be derived from further analysis of the original data.

Summarizing, the industrial category of the workplace will not provide a direct indicator of work-travel patterns, but it may include a number of characteristics that, in a given metropolitan area, will bear fairly direct relationships to travel patterns and modes selected.

[23] *Home Location Pattern of Industrial Workers in the Detroit Region,* 1955.

4.3. ANALYZING PATTERNS OF TRAVEL TO WORKPLACES

In a particular metropolitan area, it is possible to find classifiable patterns in the distributions of residences about workplaces and in the systems of travel between them. Our objective is predictability of the work-oriented systems of urban travel based upon understanding of the subject phenomena. The objective may be first sought in techniques of describing observed patterns, and in seeking out the essential elements necessary for reproduceability of such patterns. If recently existing patterns can be reproduced from the elements, then there are grounds to expect predictability by reasoned adjustments of the elements applicable in a design or forecast year.

Three types of study of pattern are suggested in the paragraphs following. All three are of the relationship between a trip-confluence center and its surrounding field of influence. They are:

1. Cumulative percentage distribution of trips from places of origin, by mile- or time-distance. These may be plotted on probability coordinates, or on log-probability coordinates as an aid to visualization of patterns (see Figure VII).

2. Study of the distribution over mile- or time-distance of the per capita proportion of trips arriving at a destination from each distance class. Logarithmic treatment of data is suggested (see Figures VIII and IX).

3. Standardization of the distributions of Item 2 above in terms of volume-scale of the destination area.

Findings of pattern by the three suggested methods, as an aid to understanding of the chorology of residences and workplaces, are summarized in Chapter VI. The remainder of this chapter is concerned with techniques of observation and analysis.

Figure VII

DISTRIBUTION OF RESIDENCES ABOUT WORKPLACES IN AREA "J", BY MILE DISTANCE PHILADELPHIA - 1947

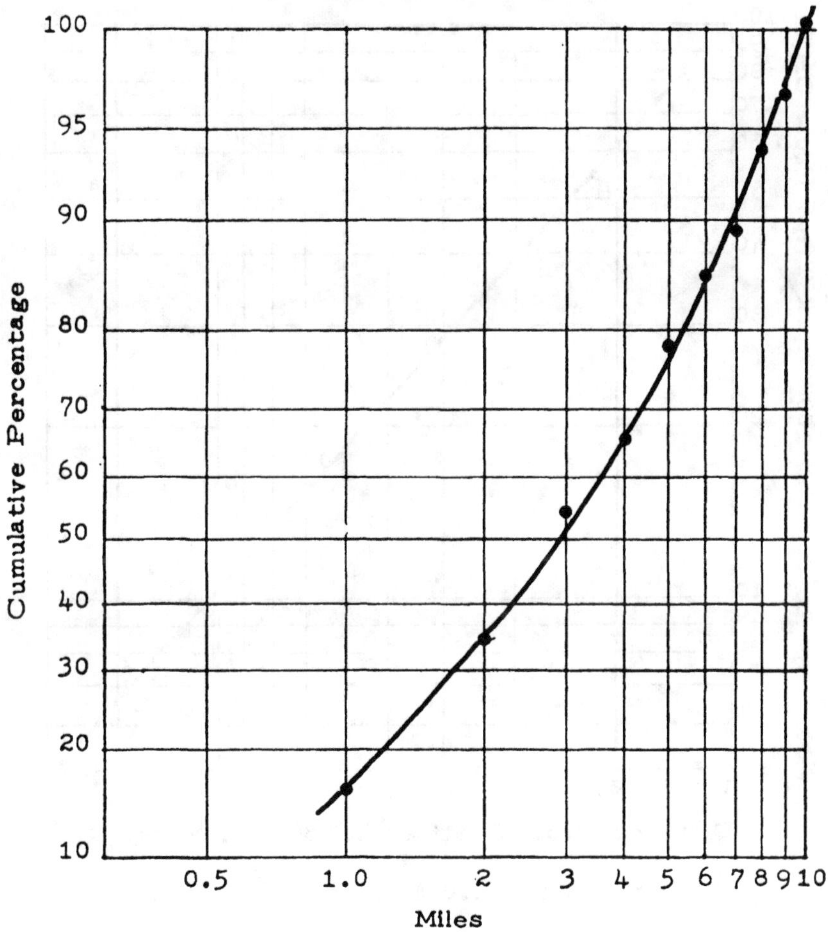

Figure VIII

PROPORTION OF ALL ORIGINATING WORK TRIPS TO 086, BY MILE DISTANCE FROM 086
PHILADELPHIA - 1947

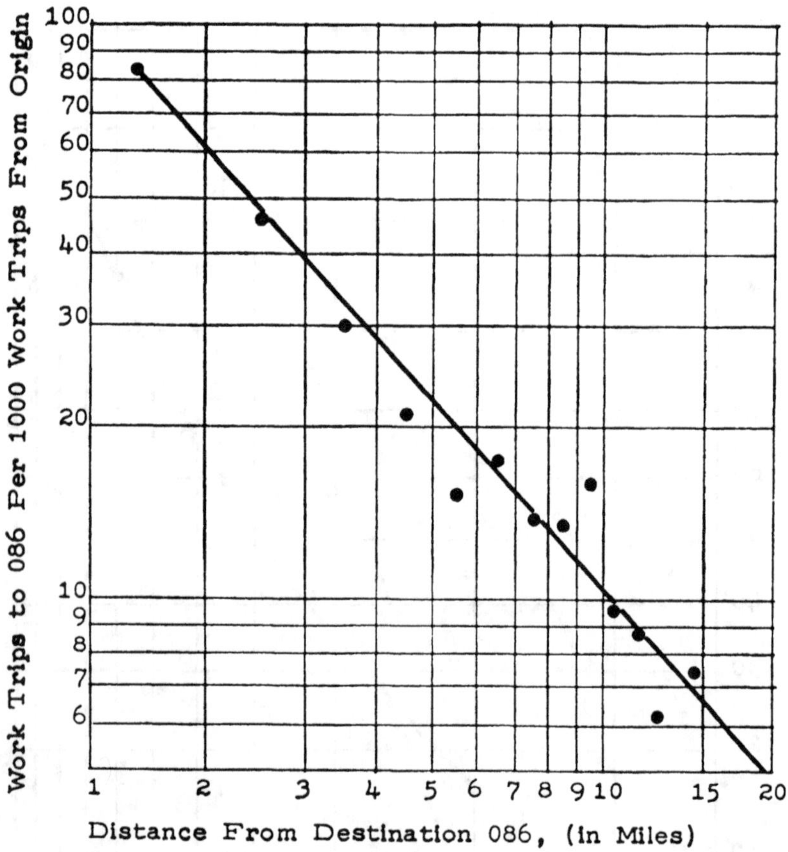

$\log y = 2.076 - 1.064 \log x \qquad \bar{r} = 0.97$

Figure IX

PROPORTION OF ALL ORIGINATING WORK TRIPS TO 037, BY MILE DISTANCE FROM 037
PHILADELPHIA - 1947

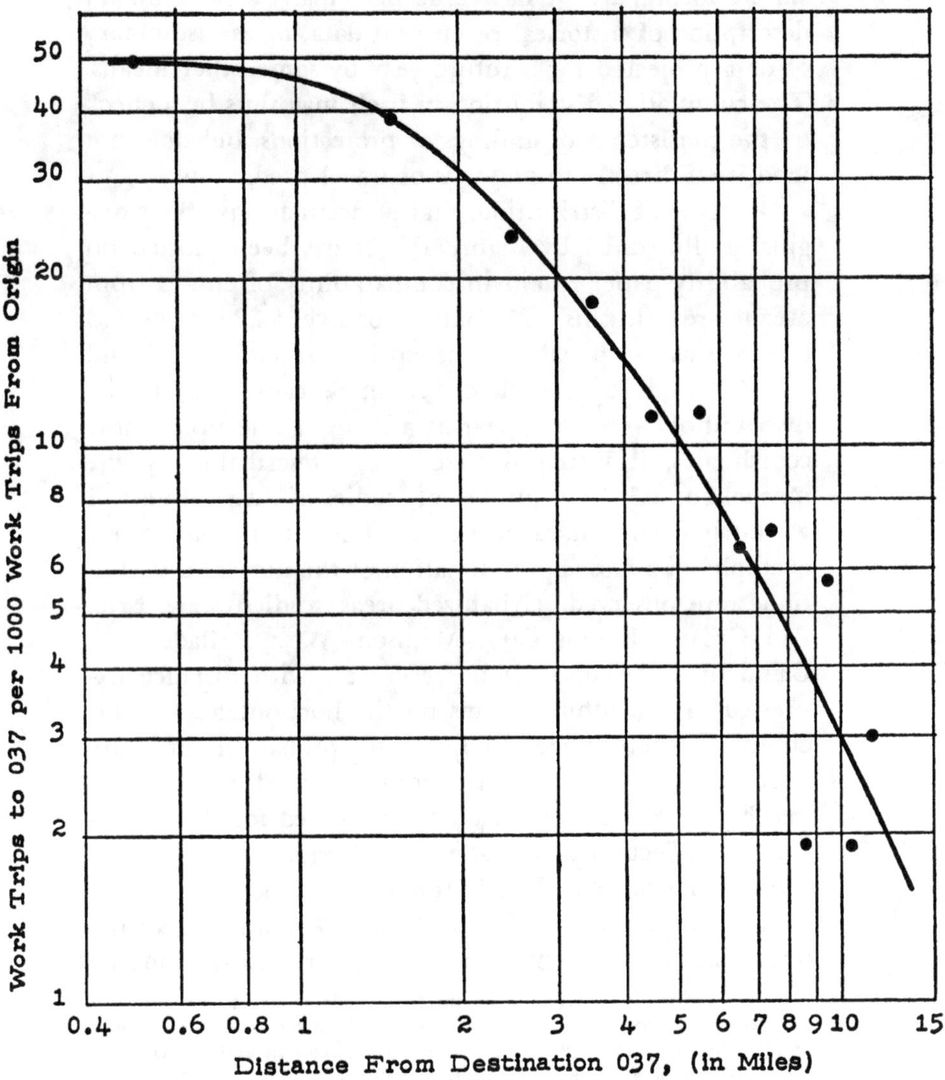

$$\log y = 1.668 - 0.341 \log x - 0.841 (\log x)^2 \qquad \bar{r} = 0.94$$

4.31. Cumulative Percentage Distributions

Study of the patterns of trips in terms of cumulative percentage distributions has the virtue of being fairly simple to prepare. It has value principally as a summary description of historical or current data, or as a summary of data projected for a future year by some other means. The cumulative distribution in itself may thus be a check on the consistency of findings or projections, but does not lend itself directly to methods of travel prediction.

One type of distribution, that of destinations about origins, is illustrated by Figure III. It has been plotted on probability paper to aid in comparability of curves from area to area. (In this case, comparison over a 22-year period of time was also possible, since equivalent data were found for 1934 and 1956, as described in Section 2.21.) If the distribution were to appear as a straight line upon these coordinates, it would represent one describable by the normal curve.[24] In a number of studies of distributions of work-trip origins about workplace destinations, only a few examples of a close approximation of the normal distribution were observed. Urbanized areas studied were Erie, Pa.; Greater Kansas City; Madison, Wis.; Philadelphia-Camden; and Tulsa, Okla. However, with distance expressed in logarithmic terms on the horizontal scale and cumulative percentage of trips on the probability ordinate scale, coherent distribution patterns obtained in most cases (see Figure VII for one example, prepared for the Olney-Oak Lane Section of Philadelphia, Section J).

Since the normal distribution of residences or origins about workplaces is not expected, it is perhaps only a curiosity that fair approximations to log-normal distributions

[24] The data are not expected to be describable in exact terms by the normal curve. Plotting the data as described here has the purpose only of providing a form of shorthand in which changes in and character of the distributions may be more readily apparent.

were observed for origins of auto work trips about several industrial workplaces in Kansas City, Mo. This was found to be the case for trips to the Blue Valley, Northeast, and Fairfax industrial districts.[25]

As was observed for national distributions of trips from origins (1941 to 1951 as shown in Figures I and II) and for areas within Philadelphia (1934 to 1956), a sense of change (and rate of change) in distributions over time may be obtained from this type of study; but a sufficient basis suitable for predictive purposes is not found. However, a glaring error in computation of future proportional distributions of trips would probably show up in a graphical distribution analysis of current and projected trip data.

This type of analysis was also prepared for trips with one terminus in the Philadelphia CBD in the year 1947 with the purpose of trip as the third variable. The systems of trips made for business, work, and shopping purposes were found to have essentially similar patterns, with differences remaining fairly constant in absolute mileage throughout the distance spectrum. The median work-trip length was found to be about 0.3 miles greater than the median business trip to a CBD destination, and about 0.3 miles shorter than the median shopping trip to the CBD.

4.32. Residences about Workplaces: Standardization of Origin Area Characteristics

In studying confluence patterns at various workplaces, in terms of distances between residences and the workplaces, it is informative to eliminate if possible the effects of varying densities in the spatial distribution of the labor

[25] Principal data source was: *A Traffic Survey of Greater Kansas City*, by the Missouri and Kansas State highway agencies, city agencies in the two Kansas Cities, and the U.S. Public Roads Administration (Bureau of Public Roads), 1944-45. Also consulted were *Expressways, Greater Kansas City*, City Plan Commission, Kansas City, Mo., 1951 (an excellent report) and *Locational Costs in Industrial Districts, Greater Kansas City*, City Plan Commission, Kansas City, Mo., 1950.

force. Standardization of the data may be achieved fairly well by considering the proportions of all originating work trips from each traffic district that have destinations at the workplace under study. These proportions can then be plotted against the distances of the individual origin districts from the respective workplaces. It may be said generally that persistent patterns are discernible for the proportions of work trips bound to a particular destination of all work trips leaving origins in each mile-distance-ring about that workplace center.

A typical example is shown in Figure VIII. The illustration shows the common straight-line pattern. The data represent 14,400 trips of 6.3 miles median length from origins about a facility for the service and repair of ships, located in South Philadelphia within traffic district 086, a unit of 1.2 square miles gross area. The data are described by the equation,

$$\log y = 2.076 - 1.064 \log x,$$

which may be rewritten as, $y = 119.1/x^{1.064}$.

The correlation coefficient was found to be high (over 0.9) and statistically significant. This curve is considered to be a prototype for the pattern in which work trips originate about a workplace with a relatively large number of employees of diverse skills within a compact area. For the largest grouping of work trips to a single focus, the special case of the CBD considered as a destination locus, the precision of measurement was found to be disappointingly low. (This condition may be solved by using smaller zonal areas.) All that may be said about the relationship, as viewed on log-log coordinates, is that within the internal area the applicable description appeared to be

$$\log y = \log b + a \cdot \log x.$$

(The exponent *a* is always negative.) The slope of the straight line was considerably less steep than that for the distribution of origins about District 086.

Gross data of the CBD study show that of the 916,000 originating work trips considered, 212,000 had destinations in the CBD. Median work-trip length to the CBD was 4.0 miles, compared with 3.8 miles for trips to the CBD for all purposes. CBD-destined work trips varied from a low proportion in the internal area of 132 per 1,000 originating from Section L (Far Northeast community) in Philadelphia, to a high of 351 per 1,000 from Section B (South Philadelphia) (See Figure IV). The attraction of the CBD as a workplace was strong throughout the internal area. The *rate* of work-trip attraction to the CBD could not hold for many more miles into the suburban, external area at the level observed. The data plotted for the internal area indicated a slope of line that, if extrapolated, would show an attraction of more than 80 per 1,000 work trips to the Philadelphia CBD at a distance of 100 miles. This extrapolation is not acceptable on the basis of what is known. It may be assumed then that the slope of the line would increase in steepness at distances beyond the internal area. It is inferred that an approximation of metropolitan trip-attraction to the CBD may be made by means of a simple power-function expression of the type

$$\log y = \log b + a \cdot \log x,$$

or,

$$y = b \cdot x^{a}$$

but that a more complete consideration would indicate a curve of greater complexity.

Within the internal survey area, examples may also be found of more complex relationships than the simple power-function (straight line on log-log coordinates). Figure IX indicates the "decay" with distance of attraction of work trips to destinations in District 037 in Philadelphia.[26]

A parabolic arc was fitted to the distribution. The decay with distance in miles of attraction of work trips to District 037 was found to be describable by the expression

$$\log y = 1.668 - 0.341 \log x - 0.841 (\log x)^2.$$

For this relationship, the coefficient of correlation was found to be over 0.9 and statistically significant.

This curve of attraction versus distance indicates that despite the low median trip-length of 2.9 miles the proportion of trips to work from the immediately surrounding area was thinned out by some factor or factors. There may have been a poor matching of the resident work force with the nature of the locally available jobs, or the location of the district on one of the major vehicular arteries of Philadelphia may have caused the apparent deficit of resident workers in local workplaces. Though the various possible factors may be listed, the problem will not be closer to solution by reason of speculation. Additional tools of analysis, accounting for effects of the relevant factors, are to be applied in our quest for understanding.

Study of work-trip attraction over distance was conducted in the manner described above for traffic districts in several cities.[27] In the Philadelphia examples, the me-

[26] District 037 occupies about 2.5 gross square miles in the north-central part of Philadelphia along Roosevelt Boulevard, a major traffic artery. In 1948, 6,706 manufacturing jobs were listed in this District. Total employment was not known. The 1947 O-D internal survey indicated about 12,500 destination work trip-ends in District 037.

[27] Study was made of a number of districts in Philadelphia, including those listed here, for which distributions in full detail were prepared: Districts 015, 033, 036, 037, 051, and 086. Ease of matching districts with census tract information was one of the criteria of selection. Other districts studied included two non-CBD districts in Toronto, 112 and 162; a peripheral district in Madison, Wis., in which is located a meat-packing plant employing over 3,000 persons, District 077; and one of the several CBD zones in Erie, Pa., 0021. The last-named was the only area for which a coherent pattern did not appear by this type of analysis.

dian lengths of work trips were observed to vary for the several destinations from 2.8 to 6.3 miles. Some of the destinations with less employment were observed to be associated with median trips of greater mile-distance than were found for the larger destination districts. (This contrasts with the findings reported for Flint previously referred to.) As stated previously, the median trip length from internal origins to the largest workplace center, the CBD, was only 4.0 miles. Thus, the median trip length appeared to be affected more by the location of the destination within the settled area than by any other noticeable factor. The sample of districts was small in number; nonetheless, with the exception of the district in the Erie CBD, the results obtained for the several destinations were internally consistent and comparable.

4.33. Residences about Workplaces: Standardization of Workplace Size Characteristics

The patterns of the spatial distribution of work-trip origins about workplaces as described in the preceding section are visually comparable in graphed form. Since the workplace destinations, however, represent varying proportions of the urban or metropolitan area's employed force, the plotted data appear as a family of curves, located at differing positions in reference to the vertical scale or ordinate values. To gain a sense of the relationship without regard to size of the areas in which trips begin or terminate, the procedure of the preceding section may be further refined by considering the proportions of work trips to a particular destination as determined above, and standardizing in terms of a measure of central tendency for that particular distribution. Examples are given in Table IV-4.

In Table IV-4, the proportions of all work trips from each mile ring (around each destination) which terminated in the two example districts (037 and 086 in Philadelphia)

TABLE IV-4

DISTANCE-VARIABLE RATIOS: WORK TRIPS TO TWO WORKPLACE DESTINATIONS IN PHILADELPHIA, 1947

Trip length (in miles)	Work trips to 037: per 1,000 work trips from origins[a]	Work trips to 037: Distance-variable ratio[b]	Work trips to 086: per 1,000 work trips from origins[a]	Work trips to 086: Distance-variable ratio[b]
0- 0.9	49.1	3.24	0.0	—
1- 1.9	38.7	2.56	83.2	5.24
2- 2.9	23.7	1.57	46.0	2.90
3- 3.9	18.0	1.19	30.0	1.89
4- 4.9	11.3	0.75	21.4	1.35
5- 5.9	11.4	0.76	15.2	0.96
6- 6.9	6.5	0.43	17.4	1.10
7- 7.9	7.0	0.46	13.7	0.86
8- 8.9	1.9	0.13	13.5	0.85
9- 9.9	5.7	0.38	16.1	1.02
10-10.9	1.9	0.13	9.5	0.60
11-11.9	3.0	0.20	8.7	0.55
12-12.9	0.0	—	6.2	0.39
13-13.9	0.0	—	0.0	—
14-14.9	0.0	—	7.4	0.47
TOTAL	15.1	1.00	15.9	1.00

[a] District 0.37. 12,500 work trip-ends in a universe of 826,300 internal work-trips origins. 1.51%; District 086. 14,400 work trip-ends in a universe of 906,000 internal work-trip origins. 1.59%. Size of deck varied from that for 037 for minor technical reasons.

[b] Distance-variable ratio derived by dividing proportion originating from each distance-ring by weighted average proportion; i.e., by 15.1 for 037 and by 15.9 for 086. Median trip-distance to 037 was 2.87 miles and to 086 was 6.34 miles.

were divided by the respective weighted average proportions (per 1,000) of all reported work trips to these destinations. The weighted arithmetic average is to be used as a divisor, rather than a figure corresponding to the trip-proportion which might be calculated as originating from a thin annular belt around the destination at the distance of the median trip length. (The latter point derives from the intrinsic meanings of the two types of measures of central tendency.)

It is relevant to indicate that the scale-adjusted data of Tables IV-4 and IV-5 may be plotted on log-log coordinates to yield curves similar to those for which algebraic relationships have been calculated in the preceding section. The division of the ordinate values by a constant has of course no effect upon the form of the curve as it appears in this type of representation. The only effect is to change the position of the curve by sliding it in a direction parallel to the vertical axis.

A simple means to describe the essential relationship between a workplace destination and the surrounding distribution of origins of work trips made to it has been discussed in these examples. This relationship has the weaknesses of the several-faceted meaning of "work trip," as described in Appendix C. Nonetheless, for purposes of building up a structure of the patterns of travel, this form of developing the data has value. For an individual workplace, such description has considerably more meaning than the usual dot-map of employee residences about the workplace. The dot-map has a visual impact showing the distribution at a particular point in time, but is of little analytic worth.[28] The possible uses of analysis by means of the "distance-variable" ratio for the study of urban transportation are discussed in Section 5.3.

4.4. ESSENTIAL PATTERNS OF CONFLUENCE

Described in Sections 4.32 and 4.33 are means for considering the relationships between deviations from the average proportion of work trips ending at a particular workplace (the distance-variable ratios of Table IV-4), and the distances to it from the origins. Now to be considered

[28] Management interested in transportation problems of the personnel have in a number of known cases prepared such maps or beaded-pin distributions to aid in establishing car pools for day-to-day or emergency transportation needs.

is a way of generalizing this information in terms of the metropolitan system of work trips.

Additionally, attention is given to mathematical expressions which appear to hold potential for usefulness in descriptions and prediction of work-trip travel patterns. Consideration in passing is given to the comparison of work-trip patterns as derived and models for description of total travel in urbanized areas.

4.41. A Matrix of the Origins and Destinations of Trips

For analysis, data for work trips originating from each district can usefully be assembled in a rectangular matrix with the corresponding districts in which they terminate. In such a matrix, the starting points appear in a column at the left of the table, and destinations appear in a row at the bottom of the table. In each cell there is a figure indicating the number of trips from a particular origin district to a particular workplace destination. For the entire matrix, a single algebraic expression may be derived to indicate the relationship to air-line length of trip of the deviations from the average proportion of trips to the destination. This concept of course may be used to develop algebraic descriptions of the systems of trips for each functional purpose, but the scope of the present effort restricts our consideration to work-purpose trips.

In paragraphs following, a description is given of a pilot study of work trips by drivers of autos (using Philadelphia data) as a single illustration of a means of preparing a matrix assembly of work-trip data. The approach indicated in the study is more significant than the numbers shown or mathematical expressions derived.[29] Even with its limi-

[29] The main limitations of the study were in the size of the area units selected and in the sampling of trip interchanges. The units were larger in area than is desirable for reasonable precision in selection of zone centroids, and the trip interchanges were selected not only for size but for

tations, however, the findings of the pilot study were indicative of the desirability of taking it from the desk-calculator stage to that of the automatic calculator. The study has been reported in the April, 1957, issue of *Traffic Quarterly*.[30]

The *Traffic Quarterly* article describes an algebraic relationship developed between ordinate values closely equivalent to what are called in this chapter "distance-variable ratios" (Table IV-4) and the time-length of work trips made by auto-drivers. The curve fitted to Philadelphia data for 20 interzonal traffic interchanges is shown in Figure 3 in the article. The ordinate values for the curve shown are indicated as Ratio: Actual to Probable Volumes. These ordinate values are derived from work-trip data for the 20 permutations[31] of two zones, hence differ from the "distance-variable ratios" which are calculated from data grouped by distance-intervals about a single destination. However, the basic concept is the same. It lies in the measurement of the ratios, at varying distances, of proportions of originating trips to a given destination to the weighted average proportion of all work trips to that destination.

In the example, a decision was made to utilize as simple a mathematical expression as was compatible with the nature of the data, in order to maximize potential applications of the approach. Hence, a curve representing a special rather than the most general case was employed. The curve fitted to the data was an equilateral (or rectangular) hyperbola, which in the formulation employed has the char-

geographic representation. This hand-picking was a source of observer bias that could have been eliminated by use of all the permutations of zonal pairs, or of a random sample of them.

[30] "The Analysis of Work-trip Data . . . ," Howard S. Lapin, pp. 278-292. Refer also on this subject to DMATS, *Data Summary and Interpretation*, Part I, pp. 92-94, July, 1955.

[31] Since a system of trips is under consideration, the term permutation rather than combination is employed, because the order of arrangement of the pairs of zones is significant; i.e., which is the origin and which the destination.

acteristic of becoming asymptotic to the x axis.[32] The curve is described by the equation $1/y = c + bx$. A more general form (discussed in the next section) is the exponential expression, $y = c + bx^{a}$. The curve in the example under discussion may be most readily plotted on equally spaced (Cartesian) coordinates. The second form lends itself quite well to plotting on log-log coordinates.

Following are observations made from the pilot study for Philadelphia. In the *Traffic Quarterly* article, it was suggested that careful consideration be given the points (or O-D interzonal traffic exchanges) which plotted at an unusual distance from the fitted curve. At least two types of study were outlined:

1. Compare the nature of the resident labor force with the nature of the jobs in the destination zone.
2. Notice quality of vehicular routes and transit service between the zones.

It was inferred that points located above the curve corresponded with a close matching of the work capabilities of the resident workers with job opportunities in the destination zone, together with at least adequate transportation service. Points located below the curve might be studied for conditions of poor matching of resident and workplace industrial characteristics, or poor transportation service, or both.

An application of the Philadelphia example was made in a preliminary study by the Philadelphia City Planning Commission at a time when an approximate but comprehensive analysis of 1980 trip distributions was urgently required.[33] In the PCPC study (prepared at a desk-calcu-

[32] The hyperbolic curve for Philadelphia data is described by the equation $1/y = 0.151 + 0.327\ x$. In the example, 74 per cent of variation in ordinate values was explainable in terms of the time-distance (abscissa) variable. The coefficient of correlation was 0.86. This footnote supersedes errata in the article, for which *Traffic Quarterly* was not responsible.

[33] Robert B. Coughlin, *op. cit.*

lator scale of fineness), located 1980 population and employment values and estimated 1980 travel times were used in preparing a gross total structure of 1980 work trips; by the use of several multiplier factors, adapted from current experience, an approximation was suggested for afternoon peak-hour automobile trips of all purposes.

4.42. General Forms

Insofar as "per capita" or "probability" analysis of confluence of work trips at destination areas is concerned, the evidence obtained indicates that the most general form of expression is indicated by the power function, $y = c + b \cdot x^a$, where a is a negative exponent, and c and b are values to be determined for a given system of trips.[34] The approach for study of a single destination was illustrated by analysis of the field of work-trip origins about District 086 in Philadelphia. Not to overstate this position, some exceptions have been noted, and an example of a distribution which does not fit into the simple general form has been given in the case of District 037 in Philadelphia.

The expression $y = c + b \cdot x^a$ has the desirable characteristic that a curve representing it becomes asymptotic to the x or distance axis, with increasing values of x. The aspect of diminishing interchange between zones with increasing physical distance is desirable in any general formulation of a communication function. Thus the mathematical format has a major characteristic compatible with the nature of the phenomena being described.

For another view of this matter, the reader is referred

[34] The expression is derived by considering the proportion of trips to a particular destination of all similar trips from origins, at each distance class about the destination. The y is a measure of such originating trips to the given destination; x is the distance measure; b is an ultimate but unattainable frequency of trips which may be measured at one unit of distance from the destination.

to an interesting document by Godlund.[35] His analysis concerns bus trips for all purposes, made from rural and suburban hinterlands to urban centers. It is to be noted that while Godlund did not control the purpose of the trips, he controlled the mode selected and the trip direction. His analysis indicates a finding of the general applicability of the function

$$y = c + b \cdot x^a \qquad \text{(Equation 1)}$$

in describing numbers of bus journeys by passengers traveling to urban destinations in Sweden. He mentions his own findings of values of the a exponent in a range from -1.5 to -2.5, and findings in other studies of all forms of traffic in Sweden in a range from -2.0 to -2.8 (see pages 51-52 in Godlund's bulletin). It should be noticed, of course, that all Godlund's work refers to distance measured in kilometers.

Important and worthwhile research conducted in Sacramento offices of the California Division of Highways bears attention.[36] Osofsky described work on trip-prediction models for three urbanized areas in California: Modesto, San Diego, and Vallejo. In order to reduce errors in synthetic reproduction of known travel patterns, different factors, exponents, and formulations were required in each case, though the general form was similar to the second arrangement of the Vallejo model, which was found to be the most successful. The Vallejo "Mark II" may be stated generally as

$$y = c_o + (c_1k_1 + c_2k_2 + c_3k_3 + c_4k_4) \cdot (1 + x)^a$$

(Equation 2).[37]

[35] Sven Godlund, *The Function and Growth of Bus Traffic within the Sphere of Urban Influence*. Bulletin 18, Department of Geography, The Royal University of Lund, Sweden (Lund, Sweden: C. W. K. Gleerup, 1956).

[36] See reference of footnote 2, this chapter.

[37] In this model, the y values represent volumes of trips between combinations of pairs of zones or between combinations of one cordon station

The exponent a is given as -2 in this statement. Equation 2 might now be restated in another and simpler form, where the term B is a multiplier calculated for a particular trip-interchange volume between two specific zones. Thus, it may be restated in the form

$$y = c_o + B \cdot (1 + x)^a \quad \text{(Equation 3).}^{38}$$

Equation 3 may be compared with the general form of the power function described in the first paragraph of this section.

Osofsky also reported on a comparison of results of the Modesto model with a gravity method formulation prepared for that city. He indicated that the regression model was established to reproduce trip interchanges more accurately than could be done with the gravity method (which employed a statement for the "friction factor" y, as $y = b \cdot x^a$, where $a = -1.63$).

The Vallejo-type model represents a highly efficient and commendable empirical hypothesis. However, it is desirable that a "kitchen-sink" regression model be also rational, in the sense that effects of inclusion or exclusion of the several variables in the B term should be explainable, either before or after a basic hypothesis is established. Meaning is desired for the inclusion of specific variables: one facet of this subject in respect to the Vallejo model is touched upon in Section 4.21.

We have a situation where at least two approaches to a

---

and one zone. Two sets of multiple regression factors must be derived for these two types of cases. The factors so determined are the c values, and are unique in each urbanized area. The k values represent the particular zonal characteristics found to give the most successful results. They include population, number of employed persons in each zone, etc. The x values represent distance between zones or stations and zones. No distinction is made for purpose, mode, or direction of trip. Trips for all purposes are included.

[38] More precisely, for the interchange between two given zones i and j, the formulation would be

$$y_{ij} = c_o + B_{ij} \cdot (1 + x_{ij})^a \quad \text{(Equation 4).}$$

problem may be taken, and each has its own advantage and disadvantages. Perhaps the two approaches are not mutually exclusive. It is suggested that the approach of the present report, the isolation of *systems* of trips by purposes, is essential to meaningful assemblies of data. At the same time, it is acknowledged that the inherent errors of the gravity formulations mentioned by Osofsky are correctly stated. It is through this recognition that error may be reduced or compensated for in the gravity formulations. In accepting the framework of gravity formulations that can be constructed from the ground up with current data, we should be able to approach more closely to the goal of developing phenomenological and rational wholistic models.

4.5. SUMMARY

A number of studies of relationships between characteristics of communities and their workplaces and the nature of the work trips serving them were discussed in Sections 4.1 and 4.2. In the study of daytime population distributions some of the pairs of factors studied were found to be closely associated, e.g., location of workplace with respect to an urban central point, and the directional distribution of trips by length of trip. Other pairs of factors had no noticeable association, e.g., location of workplace and occupancy of private vehicles in the journey to work.

Further description and analysis of patterns of travel from residences to workplaces were presented in Section 4.3. Absolute and relative-scale (or standardizing) techniques were indicated.

The assembly of available information, such as that of Section 4.3, into a general method of analysis has been discussed in 4.4. From what has been presented in the latter section, the reader may gain the suspicion that all the final answers are not yet in, so far as analysis of urban travel is concerned. He will be right.

We have followed in detail some means of describing patterns of travel. We have also seen that a synthetic model of many variables may be more precise than a gravity formulation in reproducing current travel patterns. A resolution of the two methodological approaches seems indicated through further analysis of the effects of individual variables, and investigation of concepts such as those mentioned in Section 4.1.

The ideas and techniques described in this and in preceding chapters are considered to hold valuable potential in small cities, in low-budget research programs, and in situations where study is to be directed toward a particular focus (as in the case of a proposed housing development, or an individual manufacturing plant). The approaches described hold potential also for use in large-scale research projects, largely for the understanding of operative factors which may be gained. A desirable product of large-scale, intensive research would be a general model providing a description and rational prediction of total travel patterns, incorporating studies by major systems of trips and by major foci, such as the CBD and other large workplace areas. In any case, the criteria upon which selection of technique and direction of study or research must ultimately depend are the staff and money available, as these relate to the scale of the problem area.

V

Factors of Change in Patterns of Work Travel

AVAILABLE AND MEASURABLE INFORMATION ON THE JOURNEY to work shows noticeable changes in patterns over the years. Some of these changes have been described in foregoing chapters, particularly those dealing with the most obvious characteristics of mile-distance, duration, and mode used. This chapter considers briefly several factors associated with, and possibly causative of, some of these changes. Fuller understanding of such related factors, and of the ways in which they are changing over time, is vital to any attempt at prediction of future patterns in urban travel. Principal factors considered are:

1. Geography of industry
2. Characteristics of employment with changing technology and productivity
3. Finance and technology of local travel
4. Changes in residential location

5.1. GEOGRAPHY OF INDUSTRY

There are several broad patterns in the changing geography of industry that affect urban areas. The dominant patterns appear to be decentralization nationally, centralization within broad regions (for instance, with the Middle Atlantic States considered as a region), and decentralization in metropolitan areas. Variations upon these themes may be caused, of course, by special efforts, particularly those of the wave of newly forming industrial development agencies. American industry, nevertheless, appears to be reaching a scale of output in many of its segments which permits a more market-oriented distribution of production centers among the geographic regions of the country. In these regions a location is sought within the limits of a reasonable resolution of major factors including the availability and expenses of transportation, materials, labor, etc. A central or widely accessible location within the market region, however, may be outlying with respect to the central city of an individual metropolitan area. Regional shifts of industry, as experienced at present, are not a recently initiated phenomenon. For instance, the shift of the full-fashioned hosiery industry away from Philadelphia and the Delaware Valley began in the decade of the 1920's.[1]

In addition to the effect of trends within particular industries, the changing composition of industry must be considered. The recent growth of the electronic and electrical, nucleonic, pharmaceutical, and chemical industries has brought into being many new plants with special locational requirements. A common element of these industries is their heavy orientation toward, and dependence upon, industrial research.

The spread of employment opportunities from central cities outward into suburban areas is most clearly discernible in terms of industrial or manufacturing employment. A secondary characteristic of the shifting locus of

[1] *Employment and Wages Survey of the Twentieth Century Fund*, W. S. Woytinsky, Research Director (New York: 1953).

employment is that of the creation of centers of commercial activities in suburban shopping centers. Industrial location and relocation in suburban settings have been significant phenomena in American industry since shortly before World War II. It is not within the scope of this report to examine reasons for such moves, but that this trend is operative and measurable is relevant.[2]

Another major factor is the greater emphasis in recent years upon industrial development as a part of urban and metropolitan renewal as conscious efforts are made to strengthen the basic economies of cities and metropolitan areas. This latter activity, to some extent, works to centralize the location of industry by making it possible for existing plants to remain or new plants to move to a built-up area. An important aid to the redevelopment effort is presented by the loop-and-radial systems of urban freeways being constructed throughout the country. Hence, despite a general trend toward industrial plant location in outlying areas, there are significant forces for a simultaneous countertrend to established urban centers.

5.2. CHARACTERISTICS OF EMPLOYMENT WITH CHANGING TECHNOLOGY AND PRODUCTIVITY

Two characteristics of employment in the future which will be highly relevant to the present subject will be changes in the real income of individuals and the numbers of weekly hours they will work.

The importance of continued increase in real income may be seen in recent history. As real income has increased, a greater number of persons have found the free-

[2] An early measure of this for the Chicago metropolitan area is given by Leverett S. Lyon in "Economic Problems of American Cities," *American Economic Review Supplement,* March, 1942, as reported in the Twentieth Century Fund study, *America's Needs and Resources.* J. Frederic Dewhurst and Associates, 1955, p. 495.

dom to seek new homesites, to purchase automobiles, and in the aggregate, to contribute to the low-density growth of cities outward from their peripheries. Since, in the next twenty years, gross population, productivity, and national product are all expected to continue to increase more or less as in recent years, the physical effects wrought upon urban areas with increasing real income per family unit are expected to continue.

Perhaps the effects of the trend in hours worked per week by the gainfully employed will be even more important to travel patterns. "The trend of working hours both in agricultural and in nonagricultural activities has been steadily downward over the past century, and this trend appears to have been accelerated in recent decades." [3] Over the decade of the nineteen fifties, the decline in the average workweek is estimated to be 2.5 hours for agricultural and nonagricultural pursuits combined, or an average of 0.25 hours per year. From this, it would appear that an average 32-hour workweek in the United States might be expected around the year 1980. At that time, or sooner, it seems likely that arrangements for a four-day rather than five-day workweek would have been initiated in some establishments. A widespread adoption of this 20 per cent reduction in numbers of work trips would probably represent the *coup de grâce* for any nonsubsidized transit systems that might be extant a generation hence. This does not imply that no public transit service will be provided, but rather that whatever transit service will be available would probably be dependent upon some measure of public support from tax revenues, in addition to substantial tax relief for the transit operation itself.

Further discussion of the present and future need for public transit service and the financial problems of pro-

[3] J. Frederic Dewhurst and Associates, *America's Needs and Resources, A New Survey* (New York: Twentieth Century Fund, 1955), p. 750.

viding it are beyond the scope of the current effort. The reader is referred to *Plan and Program 1955,* a publication of the Urban Traffic and Transportation Board of Philadelphia,[4] and to Wilfred Owen's book, *The Metropolitan Transportation Problem,* especially Chapter VI and pages 255-262.[5] In the latter reference there will be found not only a historical treatment of the problem of changing modes of urban travel, but also suggestions for amelioration of the effects of the change.

5.3. FINANCE AND TECHNOLOGY OF LOCAL TRAVEL

The efficiency and productivity of American industry are expected to continue to grow as in the years of the past two decades, just as the travel by persons to their workplaces is expected to become more costly and more tedious—in short, more inefficient. This paradoxical situation is conditioned by the relative incentives to investment in urban transportation facilities and in other enterprise, and by the consequent relative proportions of resources devoted to research and development of urban circulation equipment, way, and structures.

The prospects for urban circulation are not bright. The last innovation in street rail transit was made during the administration of Herbert Hoover, when at his urging a national conference of transit executives was called to evolve a modern design of a transit car. The result was the famous Presidents' Conference (or P.C.C.) Car. Development of the fairly efficient 50-passenger Diesel bus is more recent, but at this time it can hardly be considered new.

[4] In the U.T.T.B. report, see especially Chapter V: "Problems of Finance." Calculations and text of Chapter V were prepared by the writer of the present effort, in conformity with metropolitan financing principles developed by Dr. William L. C. Wheaton of the University of Pennsylvania.

[5] The Brookings Institution, Washington, D.C., 1956.

We do see some strong effort to improve the design of interurban commuter cars; this, in fact, appears to be the only significant area of development of transit vehicles.[6] On the other hand, a massive outpouring of energy and resources is devoted to the research and design of personal private vehicles and, in lesser degree, to the roads over which they travel.[7] The transit industry as a whole has been slow to innovate, not only in the design of vehicles for amenity and service, but also in other areas of market sensitivity. Few studies of the potential consumer market are available. Negative forecasts based upon head or ticket counts are not so scarce, and these have apparently influenced management policies. Other purveyors to the mass consumer markets find it necessary to go to the grass roots to find new directions for future service, and for profit. The transit industry seems to march to its own tune, on the way to its genteel grave. Exceptions to these statements may be noted in some hopeful activities in such areas as Philadelphia, San Francisco, Chicago, and Cleveland.

Part of the problem is the seemingly paradoxical fact that the price of transit riding is in many cases higher than the marginal cost of auto riding to work. At the present time, with flat fares prevailing in the transit industry and average lengths of transit rides of somewhat over three miles, we find an average price per passenger-mile of the order of six cents. The fact is that even with only one person in an automobile, the *marginal* cost of use of a "big-three" auto purchased new is five to six cents per passenger-mile. If an average automobile of one of the

[6] This statement could be proved obsolete, as the author hopes it will be, by adaptation of new power sources, such as the fuel cell, the rotary combustion engine, or the gas turbine engine, to street transit vehicles.

[7] A Galbraith image of relative allocation of resources is here invoked. See John K. Galbraith, *The Affluent Society* (Boston: Houghton Mifflin, 1958).

standard types is owned for six years, over-all cost, including fixed costs, is of the order of ten cents per mile.[8] However, since the typical family seems to own an auto regardless of whether or not it is used for work travel, only the marginal cost of travel to work is applicable for comparison with transit costs. A passenger is carried in about one out of two work trips in cities. If, as occurs in manufacturing industry, the cost of parking is absorbed by the employer, then automobile transportation competes favorably with public transportation.

Are there any hopes on the horizon that the journey to work may share in the general aspects of industrial efficiency? It cannot be denied that prior to saturation of a traffic facility, the great convenience of a modern automobile for travel from home to work over a grade-separated road presents one very compelling aspect of modernity, and for many individuals, efficiency in travel. The automobile will be a substantial part of any future travel pattern. But nonetheless, the total problem must be faced: if our cities are to have determinable form and be as livable as is within our intelligence and resources to build them, then transit service cannot be allowed to wither totally away.

What other aspects of modernity may be discerned for urban travel? There are some hopeful designs in embryonic or more mature stages, but dependent in all cases upon the economics of their application. The designs include the new light-weight suburban rail cars (such as

[8] See H. S. Lapin, "Costs of Automobile Ownership and Use in the San Francisco Bay Area; September 1, 1953," University of California, Berkeley. I.T.T.E. Technical Memorandum No. B-25, June 28, 1954. This memo primarily utilized information from the Heller Committee for Research in Social Economics, University of California. Other important references included work by Dr. Ralph A. Moyer of I.T.T.E.; Automobile Clubs of several states (members of A.A.A.); publications of Runzheimer, Dartnell, and Kelley; Planning Monograph No. 9 of the San Francisco Department of City Planning and L. Lawton's article in *Traffic Quarterly*, "Evaluating Highway Improvements on Mileage-and-Time Cost Basis," January, 1950.

the Budd Pioneer III), the Talgo train, the monorail, duorail, Alweg, and the variegated forms of rotary-wing and ducted-fan aircraft.

Insofar as conventional or near-conventional rail vehicles can be developed for use in suburban services, these can well aid desirable commuter transit facilities. Experiments in the Philadelphia area tend to confirm this assumption (particularly on the services to Media and Paoli of the Pennsylvania Railroad).[9]

Monorail (and a derivative, duorail) has seemed for many years to be a potential technical solution to problems of suburban and intercity surface travel. However, insufficient capital has been available to bring this type of vehicle to operational level at speeds necessary for competition with the automobiles. A few limited monorail operations have been started in this country, usually for service at fairs or expositions, but the great problems in developing feasible technical devices for a major system still appear far from solution. Among these are switching, overcoming sway in the vehicle as it approaches the stations, developing a suitable overhead structure to absorb the loads of acceleration and deceleration of a multicar train from speeds of 70 or more miles an hour. An actual intercity monorail with two-car trains (the "Schwebebahn"), in use in the Wuppertal Valley of the Ruhr, Germany, operates at an average speed of 18 mph on a

[9] The more publicized experiments, Operations Northwest and Northeast, do not include the use of new equipment, but involve only changes toward more attractive fares and service schedules. In the "Operations," the Pennsylvania Railroad, the Reading Railroad, and the Philadelphia Transportation Company (PTC), have cooperated with the City of Philadelphia in experiments in which the City guarantees the railroads a limited sum of potential marginal losses incurred by the added passenger benefits. Operation Northwest was commenced in September, 1958, and Operation Northeast in Philadelphia was begun in September, 1959. A public corporation to assume the city's part in the cooperative ventures began operations in January, 1961. It is known as the Passenger Service Improvement Corporation.

track supported by cumbersome A-frame structures. Maximum speed is 25 mph.[10] The monorail structural facilities are by no means the delicate lamp poles which appear in the promotional drawings. A sturdy structure would be required to absorb the loads imposed on the structure along the line of the direction of motion of the car and transverse to that direction. The major problems of the suspended monorail have yet to be worked out for an operational transit system at speeds approaching (let alone exceeding) those which can be obtained with conventional rail transit vehicles.

One highly significant effort in the development of transit vehicles is that being conducted at Cologne, Germany, by the Alweg group. This is a special group financed by Axel Wennergren of Sweden with a rail-vehicle which straddles a concrete wall and rolls on wheels located on top and on both sides of the wall. Potential speeds have been reported to be well over 100 miles an hour. We may expect to hear more of the monorail and Alweg vehicles. Still another effort, at Chateauroux, France, is to be conducted by a combination of more than a dozen French companies.

Within the foreseeable future, it is unlikely that the helicopter will be of service to the commuter mass market. Seat-mile prices of 7 to 10 cents may be feasible within the next 10 years. Even so, it would not bring rotary-wing aircraft within range of the day-in day-out mass commuting demands. Within the next decade this type of aircraft is likely to be suitable for air taxi and charter service, intercity travel in the 50-to-250-mile range, airport connections, and transport of air freight with a high value per pound. Even in the next two decades, however, though a *technical* possibility, use of rotorcraft in the mass commuter market

[10] James F. Holly, "Wuppertal Monorail," *Military Engineer,* XLII (January-February, 1950), 8-12.

is expected to be a negligible part of total traffic volumes.[11]

Another category of vehicle of possible interest in the distant future is the ducted-fan aircraft, which flies a matter of inches (or fractions of an inch) above the earth's surface, upon a generated cushion of air. Present types in development include a private passenger auto (Curtiss-Wright), and a truck-type vehicle (the Hovercraft), which has recently skimmed across the English Channel. Perhaps a bus prototype of this surface-dependent air vehicle may be expected someday.

5.4. CHANGES IN RESIDENTIAL LOCATION

National trends in population settlement may be described as centripetal with respect to metropolitan areas, considered as entities. But within the metropolitan areas, trends in the spatial distribution of population have been largely centrifugal (literally, fleeing from the center), with a continuing spread of low-density residential areas outward from the peripheries of urbanized areas. There are some noticeable countertrends of indeterminate size within the latter movement, as high-rise residential areas grow in or near the centers of major cities; witness the development in Manhattan East and central Philadelphia, and recent proposals for the central area of London.

In regard to these changes in the physical form of urbanized areas over time, it is of interest here to examine specifically the relationship between locations of residences and of workplaces as a conditioning agent in the urban "magnetic" phenomena of simultaneously attracting and repelling diverse particles.

[11] This point is elaborated in detail in a report on heliport planning submitted to the Philadelphia City Planning Commission in December 1959. The present writer was responsible for evaluation of the rotorcraft market through 1980 insofar as it concerned heliport planning in Philadelphia and the surrounding region.

Forces underlying the centrifugal movement of population from the central parts of an urban area were studied by Dewey as they operated in a limited area of influence around the City of Milwaukee.[12] Changes of residence were made toward living areas considered more amenable because of the greater availability of space. So long as the primary wage-earner retained the same employment, the changes were generally made at the expense of longer and perhaps more arduous work-journeys. Dewey also observed the effects of transportation facilities on outward relocations. He found that rapid transit lines continued to have influence in land development, but that a more widespread role was taken by the major highway facilities. About 80 per cent of the dwelling structures in rural areas of Milwaukee County were found to be located within one-fourth mile of a major county or state highway. Similarly, the areas of lowest population density were those farthest from the principal vehicular routes. It can be concluded from this study that quality of housing, local intrinsic convenience, and social climate were considered more important in the location of the home than arduousness of the journey to work.

The journey to work appears significant in the selection of a residential site primarily in setting an outside limit to the distance between home and work. The "principle of least effort," as put forth by Zipf, and minimization of work-trip length, as investigated by Carroll, still holds validity but, depending upon income levels and aspirations, tends to operate in conflict with status and other social needs. In this situation, relative space amenity, monthly payments or rental, and the character of the neighborhood and schools become major factors in de-

[12] Richard S. Dewey, *Peripheral Expansion of Population in Metropolitan Milwaukee* (Unpublished doctoral thesis, University of Wisconsin, 1946).

cisions on residential location. In recent years, it appears that the middle-high income "luxury" of the long commutation trip is filtering down through many more income levels. Thus, as real income rises, the long work-trip formerly associated only with upper-middle income groups becomes more representative of the entire metropolitan population.

A study of the characteristics of the journey to work of recent movers in the Philadelphia area as registered in the residential interviews of the National Housing Inventory of 1956, indicated the following (according to its author):[18]

1. A change in residence usually results in a trip to work of equal or greater length than that associated with the prior residence location.

2. Less than 1 per cent of the movers changed their places of employment as a result of a residential move, but over 7 per cent of the movers in the Philadelphia Standard Metropolitan Area (eight counties) moved their residences in response to a change in job location.

Concerning those who expressed the intention to move, he states:

1. Those working in the suburbs are slightly more accepting of residence moves than those working in the central city of the metropolitan area.

2. In general, the city worker does not intend to move to the neighborhood of his workplace, while the suburban worker has indicated a decided preference to move to or near the same area in which he works.

The similarity in the general findings for the Milwaukee and Philadelphia metropolitan areas tends to confirm the indications of simultaneous shortening of work trips by some and lengthening of trips by others. Residential amen-

[18] Louis K. Loewenstein, *The Journey to Work in Philadelphia: Its Patterns and Perspectives.* Unpublished paper, Department of City Planning, University of Pennsylvania, 1960.

ity appears to be somewhat more highly valued than ease of work travel for those who move to the suburbs, but for those already in the suburbs, easing the stresses of work travel is found to have been given a higher rating.

5.5. SUMMARY

Prediction of urban travel patterns calls for consideration of those factors which are likely to be associated with or possibly causative of some of the future changes in characteristics of urban circulation and, particularly, of the journey to work.

Broad trends over time in the spatial distribution of population at places of residence have been considered and the currents in industrial geography have been discussed, in order to give clues for understanding of the problems with which we are dealing. Again, as a key to what we might expect in terms of urban transportation, some information has been given about the effects on local travel services of changing industrial technology, and of apparent aspects of the finance and technology of metropolitan circulation.

More than thirty years ago, Robert E. Park described the shifting and balance of forces which produce physical changes within metropolitan areas.[14] He pointed out that the multiplicity of opportunity for labor within a city, when accompanied by adequate resources for circulation and mobility, constitutes a major force for conditioning the cultural landscape. Similarly, he stated that the seeking of locational advantages by commerce and industry, as affected by the availability of local labor, is a co-variant with home relocations, and serves also to condition the

[14] "The City: Suggestions for the Investigation of Human Behavior in the Urban Environment" in *The City*, 1925, pp. 1-46, as reprinted in Hatt and Reiss, *Reader in Urban Sociology* (Glencoe: The Free Press, 1951).

regroupings of residence in a continual interplay of cause and effect. This view of the forces determining urban or metropolitan structure is valuable in gaining a perspective of transformations in city form over time as the home-to-work relationship is changed.

VI

Suggestions and Conclusions

THIS FINAL CHAPTER PRESENTS SUGGESTIONS AND CONCLUSIONS related to the theme of the report, rather than attempting to recapitulate or summarize. Briefly, the matter covered is the following:

1. Some assumptions of traffic surveys
   The necessary assumption of repetitiveness of urban travel behavior.
2. A problem in forecasting trip production
   The question of how long into the future automobile ownership may be found the most important factor associated with production of urban trips.
3. Travel patterns
   Conclusions based upon study of graphical and algebraic patterns of work travel.
4. Analysis and planning techniques
   Suggestions for incorporating the present work into the analysis and planning techniques of traffic engineering and city planning agencies.
5. Needed data accumulation and research
   Indications based upon the work described in this report, not a comprehensive listing for the entire field.

6. Urban form

   Phenomenon of urban sprawl, and general descriptions of density and distance relationships in desirable forms of the future metropolitan area.

It should be noticed that following this chapter are a glossary of terms, three appendixes, and a selected bibliography. These items represent a greater depth in presentation of detail than was considered desirable to include in the main text. The appendixes concern:

A. Locating districts of employment in cities
B. Estimating worker interchange among cities
C. Work-trip data: sources, terminology, precision, and completeness.

6.1. SOME ASSUMPTIONS OF TRAFFIC SURVEYS

Discernment of pattern is essential in any attempt at prediction of a series of future events based upon similar past events. One of the important findings which lends credibility to formulations of future traffic volumes is that of the repetitive nature of urban travel patterns. One study has been made which indicates that the basic pattern of urban travel has a cyclical period of one week.[1] The pattern "largely, but not entirely, repeats itself during succeeding weeks." That repetition is not entirely exact suggests that the purposes of trips tend to vary from week to week and some new routes are followed, while gross volumes of travel tend to remain fairly constant on a short-run basis.

The particular study referred to is based on travel diaries maintained by 638 residents, representing 256 separate family units, of Linn County, Iowa, for 30 consecutive

[1] Traffic Audit Bureau, Inc., *Coverage, Repetition and Impact Provided by Poster Showings,* 60 East 42nd Street, New York 17, p. 24.

On this point, a profitable comparison might be made with Rannells' concept of "routines," *op. cit.,* Chap. II.

days each. In this period, all trips out-of-doors were recorded, along with the purposes for which they were made. The clear value of this small study in increasing the understanding of the motives for trips indicates the desirability of extending this panel type of study to larger cities.

On the question of the variability found in purposes for which people leave their homes, it appeared that proportionately more routine trips, to work and to shop, were made under conditions of unfavorable weather and relatively more social and recreational trips were made under favorable conditions. While this would seem obvious if the total numbers of trips made each week also varied, the approximate week-to-week constancy of total numbers of trips generated points up the significance of this information in forecasting the origins and destinations of trips.

Repetitiveness of travel patterns is the underlying assumption of all major traffic surveys and minor traffic studies. Virtually all major traffic surveys utilize the concept of the "average day." The need exists, therefore, to measure trip data over a sufficently long period of time to incorporate all relevant cycles, and thereby to minimize error in determining the "average day."

The example can be given of attempting to predict annual travel patterns on approach roads to recreation areas based upon a daily or even a weekly traffic count at any given time of the year.[2] The likelihood would be that the annual volume would not be correct because of the great variability of daily, weekly, and seasonal travel patterns on such roads. Granted that trips to work in the ordinary pattern of urban travel lend a degree of stability

[2] The reference is from the author's unpublished research report, "The Problem of Recreation Approach Roads in California." University of California, Berkeley. Institute of Transportation and Traffic Engineering, 1950. (Prepared in partial fulfillment of requirements for the degree Master of Engineering).

from day to day, patterns of travel for trips or other purposes may not be so stabilized on a weekly basis. In fact the latter groups of trips are those for which adjustments of plus 10 or 20 per cent must be introduced in utilization of O-D data.

It may well be that errors are ironed out by the fact that origin-destination interviewing may consume periods of three or four months, during which a number of days and weeks are covered. However, the point is not one to be dismissed on inspection. Prediction based upon repetitiveness is dependent upon a stable basic cycle. The problem requires investigation for the following data:

1. Relation of any single day to a truly average day.
2. Further analysis of the weekly cycle, and variations from week to week.
3. Seasonal factors in traffic measurement.[3]

It may seem that there is an incongruity in presenting the case for further study of the repetitiveness of travel in a report on the journey to work, for surely work travel is the most repetitive form of circulation in cities. It is also a fact that work travel is the most completely reported type of trip by purpose. But it may be inferred that further understanding of the cycles of travel of trips for purposes other than work will lead to better factors of adjustment for the nonwork trips than the adjustments of plus 10 or 20 per cent which are frequently used.

6.2. A PROBLEM IN FORECASTING TRIP PRODUCTION

Considered here is the question of how long into the future automobile ownership may be found the most important factor associated with the production of urban

[3] Work of the Port of New York Authority appears to be pointing the direction for seasonal analyses. See Warren Lovejoy's paper presented at the Highway Research Board, January 1959, and published in HRB Bulletin 224.

trips. We know that for studies made in past years, the highest coefficient of correlation for a single factor associated with trip production is found between trips per dwelling unit and the extent of car ownership. Thus, we have been able partially to explain data of past travel behavior. But what we are seeking are *predictors* or methods of estimating future travel patterns for as many as 20 or 25 years from the date at which the traffic counts are made. What we need to ask, before accepting automobile ownership as the chief factor for determining design of the future suburban highways and freeways, is this: Have we found a relationship among variables that describes basic behavior in urban travel? Have we found only a high degree of correlation which at this particular point in time might represent an historical accident? What will be the case when automobile ownership reaches a more nearly saturated level? The rate of decrease in the persons-per-auto ratio must "level off" at some time before there is a one-to-one relationship between the adult population and numbers of autos. At least in one large western county, such a saturation condition is almost in sight at the present time.

It is clear that the correlation mentioned is not meaningless, though in an expanding economy, it is possible to find many meaningless correlations with high coefficients simply because the items studied are by-products of the same economy. Such is *not* the case with the high correlation found between trips per dwelling unit and automobile ownership. Nonetheless, we have no proof that it is not a condition particular to this period of development in urban transportation. Study of data for one city over a period of seven years indicates that there is some stability over time in the predictive factor of auto ownership.[4] In both

[4] The studies referred to are: "A Study of Factors Related to Urban Travel," William L. Mertz, and Lamelle B. Hamner, *Public Roads,*

Bureau of Public Roads studies, automobile ownership had the highest correlation coefficient with trip production of any single factor studied. In the paper by Sharpe, Hansen and Hamner, however, it was recognized that automobile ownership "may only be a transient predictor of person trips."

In the situation in which a seven-year span between O-D surveys in a major urbanized area is the best we have, it has appeared that the extent of automobile ownership has fair stability as a predictor of numbers of trips. This particular part of the post-World War II period is not, however, likely to be comparable to the years following 1955. To illustrate, Philadelphia auto registrations increased from 231,000 in 1928 to only 257,000 in 1947, but from 277,000 in 1948 to 456,000 in 1955. The period 1948-1955, in Philadelphia at least, represents a period of change in per capita automobile ownership unlike any time in the previous 20 years, and quite unlikely to be comparable to any period of accelerated ownership rate in the next 20 years.

We have no certainty of the long-term stability of automobile ownership as a trip predictor. We know that it helps significantly in explaining current travel but the fact remains that auto ownership is itself an effect rather than a cause. We may be dealing with two effects of the same cause, and the cause may change, or the relationship between the two effects may change. If we could seek closer to the causes of the propensity to own autos and to travel in cities, then we would have surer bases for 20- or 25-year forecasts.

In the situation described, it would seem desirable to do these things:

April, 1957, pp. 170-174; and "Factors Affecting the Trip Generation of Land Use Areas," by Gordon B. Sharpe, Walter G. Hansen, and Lamelle B. Hamner. A paper presented at the Annual Conference of the Highway Research Board in January, 1958.

1. Carry out the recommendations of Sharpe, Hansen, and Hamner in obtaining on a regular and continuing basis "automobile or registration records—within urban and metropolitan areas on a basis of statistical areas such as census tracts, police precincts, postal zones, school districts, or O and D zonal areas."

2. Accept findings for the period 1948-1955 as the best we have, but attempt to extend the period of study of stability in the auto-ownership index. Second post-war O-D studies in cities in addition to Washington, D.C. and Tulsa would be the first sources of such information.

3. Attempt to find basic motivations in urban travel, possibly by the techniques now employed in market-oriented studies of consumer behavior. The present index is given in terms of autos; the need would seem to be to develop *people*-related indexes by seeking further to identify the people who travel, and *why* they travel.

6.3. TRAVEL PATTERNS

Some general statements may be made upon the bases of observed travel patterns, and their graphical and algebraic descriptions presented in preceding chapters.

In Chapters III and IV there are descriptions of regular patterns which may be found for distributions of workplace destinations about zones of origin, and conversely, of distributions of origins of work trips about workplace centers of confluence. These were studies of the relationships between a trip-origin or trip-confluence center and its surrounding field of influence. In addition, scale-adjusted or per capita patterns of attraction by destinations were described, as an introduction to the gravity concept of trip distributions. Graphical and algebraic approaches were employed in the quest for basic patterns. Some inferences may be made from the work discussed:

1. There was apparent consistency in the decay in numbers of workers commuting to workplaces in a given city with each mile-ring of increasing distance.

2. The number of employees at the workplace affects the absolute numbers of workers drawn from each distance ring, but the essential forms of distribution curves in a metropolitan area at a particular time bear similarity. Factors affecting the patterns include the location of the workplace with respect to the CBD and the type of economic activity, particularly as it affects the composition of the work force.

3. Similar distribution curves indicate similarities in probability of per capita attraction of workers from respective mile-rings. However, absolute numbers of workers drawn from very long distances are greater for the CBD's and other high-volume workplaces.

4. Patterns in the percentile distribution of time-lengths of work trips from residential areas relate roughly but consistently with the area's net population density and its distance from city center. This conclusion is based upon graphical analysis of Philadelphia data for heads of households. Over the period 1934 to 1956 the same data indicate both shorter and longer time-distances of work trips for increasing proportions of workers, with a fairly constant median length and longer average length of trip.

5. Persistent patterns are discernible for the proportions of work trips bound to a particular destination of all work trips leaving origins in each mile-distance-ring about that center.

6. Location of employment destinations in relation to the city center and the job density of the destination area may be relevant to the form of the distribution function in Statement 5, but neither factor has been adequately tested.

The first four statements hold interest for their descriptive value relative to housing and labor markets. They aid in statistical description of the volumes of persons traveling varying distances between homes and workplaces.

The fifth statement has the greatest analytical value for its applicability in synthesizing volumes of work travel from a total residence gravity field.

The first two statements bear resemblance to the study of field potential made in social physics. Essentially what is being sought is the relationship between a workplace trip confluence area and the field over which it draws trips. Thus, based upon the work of Stewart, it would seem reasonable to consider the field of employment as one of varying density of potential workers at an individual plant,[5] and generalized "equipotential" contours might be plotted about centers of employment.

The gravity concept of trip interchange to which Statement 5 is applicable holds promise for improving the predictability of systems of urban travel. It is not unique, however, as an organizing concept for synthetic travel models and, as Osofsky has pointed out, gravity models have some inherent error. Adaptation and refinement are to be sought, as described in Chapter IV and in the next section.

6.4. ANALYSIS AND PLANNING TECHNIQUES

In this section, some suggestions are presented for incorporating the types of studies and analyses described in the foregoing text into the present analytical methodology used by traffic engineering and city planning agencies.

Consideration of three factors may be stated as basic to description and detailed analysis of urban transportation patterns.

[5] John Q. Stewart, "Demographic Gravitation: Evidence and Application," *Sociometry*, XI (February 1948). See references to other articles by Stewart listed by Carrothers in the article of footnote 10, Chapter I.

1. Distribution of trip-ends at places of origin.
2. Distribution of trip-ends at places of destination.
3. Distribution of trips among places of origin and destination.

Until recently, urban traffic analysis consisted primarily of study of the third item above, the distribution of trips only. Lately some attention has been focused upon the first aspect, the origin of trips from residential areas. Special characteristics of terminal trip-ends have been given only sporadic attention, most notably in consideration of the special case of those in the central business district.

As analytic techniques improve, fairly complete synthetic models of past and current systems of travel appear possible. Means are to be sought to explain changes between the older and the more current descriptive models. As such changes can be related to independent measures of social, economic, and urban structural change over the intervening years, means will be found for developing synthetic models of future transportation behavior. The work described in this report is largely concerned with the useful description of transportation patterns at single points in recent time. From such description usable predictive techniques may be developed, as travel data for more than a single point in time become available for individual cities.

A current practice in forecasting travel patterns is to multiply current trip-interchange data by various factors of expansion to obtain the relationships among future trip origins and trip terminations. This practice is expedient and, in fact, necessary at present, but contains some bias toward a perpetuation of existing travel patterns in future plans. As clearer characteristic patterns can be found and described about the centers of dispersion and confluence, the possibility for synthetic and, thus, less biased models of future travel will be increased. The most fertile field for seeking such patterns is considered to be in the journey to work, for the reasons of scale, completeness and importance to facility planning described in Chapter III.

In the development of techniques of synthesis, graphical studies may provide clues leading toward more sophisticated analyses, and may well provide a means for testing consistency and completeness of models of trip generation and distribution. Such studies have additional value in their greater likelihood of usage in local applications as compared with abstruse formulations to which research time but not applied field time can be given (other than in the exceptional examples of multimillion dollar survey and planning efforts). More importantly though, the graphical analyses can lead the way toward generalization in algebraic formulation. Where an indication of consistency of pattern can be found and a logical basis exists for selection of a particular type of equation, then further and more detailed study by researchers is warranted.

In the long-term development of this type of analysis, mathematical gravity concepts of trip interchange hold promise for improving the predictability of systems of urban travel. Problems in applying this formulation reside in determining multiplier constants, weights for individuals, and exponents. But whether or not a solution is sought in terms of varying negative exponents of distance[6] or of weights for individuals, the function of intervening distance, $f(d_{ij})$, is not to be interpreted as necessarily the same function throughout the metropolitan field. The function of distance may be variable over the metropolitan field depending upon such factors as the purpose of trips and the location of at least one of the two zones of origin or destination in each combination or permutation.

By way of furnishing a cautionary note in regard to selection of algebraic formulations for predictive models, it must be remembered that it is important to seek a math-

[6] Carrothers, *op. cit.*

ematical form which is compatible with the phenomena of human behavior being studied. Richard J. Foote has made a penetrating critique of the potential and the dangers in regression analyses.[7] Mr. Foote refers to regression analyses made by agricultural economists in the 1920's and early 1930's, in which great emphasis was placed upon obtaining high multiple correlation coefficients in prediction. Foote called for seeking out "structural relationships." The structural relationship, by Foote's definition, is one which describes the *basic* behavior pattern in the subject being studied. Thus, in economic studies, demand curves and supply curves describe, and relations among prices at various levels in the marketing systems represent, structural relations. "At times, however, a high degree of correlation exists among certain economic variables that merely reflects, in a sense, an historical accident." Foote points out that more nearly structural relations are obtained when "variables that relate to quantities are expressed in per capita terms and those that relate to values are deflated, even though lower multiple correlations frequently are obtained."

6.5. NEEDED DATA ACCUMULATION AND RESEARCH

Several types of data are described below which the previous text or related work has indicated as desirable or necessary for further study of urban travel systems. The types of needed information are not intended to represent any kind of a complete checklist. For a broader approach to needed forms of data, the reader should consult the excellent materials prepared by the National Committee on Urban Transportation, obtainable from the Public Ad-

[7] "Avoiding Nonsense Correlations in Time Series Analysis." A paper presented at the Business and Economic Statistics Section of the American Statistical Association, September 11, 1957.

ministration Service (1313 East 60th Street, Chicago 37).

A major need in origin-destination travel surveys is for separate development and presentation of the data for each of the systems of work travel.[8] This refinement will aid in trip-end analysis and in the study of completeness of reporting by increasing the comparability of data on employment and the residential labor force with information on production of trips.

Another helpful source of needed information would result from inclusion within the decennial or special census surveys of questions concerning area of workplace and time and mode of travel.

In cities where the building inspection program is maintained on a current basis by a consistent program of residential interviews, it may be found possible to include a question or two on the workplace of wage earners to help in developing and maintaining a current file on journey-to-work patterns. The data so compiled would have greater analytic value if the city area in mapped representation were divided into a number of planning analysis areas whose boundaries were regarded as cordon lines. Thus, with a continuous traffic counting program, it could become possible eventually to relate current traffic counts to information on generation of trips. Given this degree of understanding of city-wide trip generation, planning for future anticipated needs could attain new levels of closeness to actuality. Similarly, better dealing with the problem situations in dense traffic areas might be made possible as control could be moved back closer to trip origins. Controlling traffic problems at the point of maximum conflict is essentially an after-the-fact approach.[9] The analogy

[8] The systems referred to are those from residences to workplaces, from workplaces to residences, and from nonresidential locations to workplaces. They are described in detail in Appendix C, Section 2.

[9] Compare the three stages of traffic control suggested by Mitchell and Rapkin, *op. cit.*, pp. 4-7.

might be made to the local check-dam approach to flood control as the sources of ground water in the watershed areas become known, in contrast with expensive high-dam control at a point of great depth.

Careful studies of the numbers of trip origins in relation to residence indicate the status of auto-ownership levels as the principal factor for explanation of variation in current total trip generation per dwelling unit (Section 6.2). In the immediate future, as suggested by Sharpe, Hansen, and Hamner, it would appear highly desirable to improve the available information on the distribution of auto ownership to aid in forecasting urban travel. This form of ownership is completely regulated by state governments. It should not be impossible to establish and maintain in local transportation planning offices current information on total vehicle ownership by type of vehicle and by census tract of residence or place of business of the registered owner.

Another direction in the gathering of data and in research is suggested. A number of indicators are available to show quantities of travel by the several modes in all United States cities. Compilation of annual indexes of total urban travel would be helpful in travel prediction and in gauging the long-term efficacy of metropolitan land-use arrangements. This matter is described further in the *AIP Journal* article listed below.[10]

6.6. URBAN FORM

In order to give proper meaning to previous statements concerning the need to study the journey to work for its effect upon population and employment densities, brief discussion is given in the concluding pages to the form of

[10] H. S. Lapin, "Improving the Predictability of Vehicular Travel Requirements in Cities," Technical Memorandum B-27. University of California, Berkeley, I.T.T.E., September 27, 1954. And, "Index Numbers for Urban Travel," *AIP Journal,* XXVII (August, 1961), 215-217.

cities, the widespread phenomenon of "urban sprawl," and to a goal concerning the relationship of circulation facilities and central and outlying population settlement and densities.

The spread of suburbs with a low population density around our major cities, primarily in the past two decades, has been caricatured as "urban sprawl."[11] Roughly, the following are the two current extremes in views of sprawl:

1. Sprawl is a natural result of a growing population and an expanding economy. The latter produces more buying power and new means of transportation, so that even families with an average income can afford to live farther from their work, in generally more spacious accommodations.

2. Sprawl is an undesirable, or at least uneconomic, trend. It makes the core of the urban area less important and forces people to travel farther to work. It increases the cost of municipal services and utilities. It should therefore be controlled and minimized or diverted to other physical forms.

Both views may be seen to have some merit. Further, people on both sides agree on some things. Nearly everyone agrees, for instance, that some kind of subdivision control is in the public interest, to prevent substandard housing and to help guarantee adequate utilities, shopping areas, and space for public facilities. Again, nearly everyone agrees that even though better transportation facilitates sprawl, better transportation is needed. There are, of course, a variety of views as to what kind of transportation should be encouraged.[12]

[11] Urban sprawl may be defined as the physical expansion of urbanized areas outward from the peripheries of existing development under conditions of inadequate control.

[12] See "Integrating Urban Expressway Planning Efforts," *Traffic Engineering*, May, 1956, for a brief discussion of the role of highways in relation to metropolitan resettlement.

Some planners postulate the development of satellite towns around the city, visualizing belts of park lands, farmlands, or park-expressways to separate the satellite towns from the city. In this way, it is believed, a desirable combination of big city and village centers with efficient government would be realized. One feature of this idea is that the suburban fringe residents could more readily be held accountable for the costs and privileges of urban dwelling. Some city planners have urged the development of independent new towns, or garden cities, as a means of counteracting sprawl.

One of the broadest views was that expressed by the late Charles E. Merriam, who was Professor Emeritus of Political Science at the University of Chicago. His thoughts are presented verbatim as they cannot be better stated.

> I do not anticipate the decline of urban population or the wholesale dispersion of great centers of population. A city expands but does not often disintegrate and disappear. If the urban community has a fundamentally sound economic base and has a site whose disadvantages are not too costly to overcome, then the realistic answer to the question of a desirable urban environment lies not in wholesale dispersion, but in the judicious reshaping of the urban community and region by systematic development and redevelopment in accordance with forward-looking and intelligent plans. It is possible to loosen up the central areas of congestion and create a more decentralized metropolitan pattern. Such a moderately centralized, reintegrated urban structure should have greater stability and might offer economies in production, larger public facilities and services, and make possible the attainment of higher living standards. Urban environment has no meaning except as it is worked out in terms of the daily life of the human beings who make the city what it is.
>
> . . . Anyone who follows the arguments and realizes the importance of the subject he is dealing with soon loses the dogmatism that has been all too prevalent in discussions of

this subject. Undoubtedly, we ought to have many more experiments and trials of urban groupings. The present trend in industrial relocation may indicate that the larger cities over the next generation or so may lose some parts of the industrial employment that they have had in the past. But this is a long way from the glib conclusion that the big cities are doomed and that the only intelligent thing to do is plug for a brand new pattern or theory of urban dispersal and disintegration.[13]

Dr. Merriam's conclusions impose responsibility upon transportation and city planners. The goal thus should not be the blind promotion of a particular brand or size of garden city, or formless sprawl, or the *status quo* at whatever point we find ourselves, or the meek acceptance of graphically projected trends. What is needed are individual, technical solutions which neither abandon the central city of a metropolitan area nor enshrine it in its present form for all time.

A hopeful view of the future is given by Rannells: "The coming pattern of urbanization appears to be one dominated by a network of strong concentrations tied together by transportation lines, not the general suburban sprawl that was formerly predicted." [14] Rannells implies that this is the premise upon which metropolitan planning should be based. Another similar idea is expressed by Blumenfeld, who calls for differentiated suburbs of moderately high population densities around the central city, with good access to the center and to the open country.[15] The writer emphasizes the objective that a metropolitan solution be sought which not only recognizes the factor of potential

[13] In "Problems of Reorganizing Our Great Cities." *Planning 1948*. Proceedings of the National Planning Conference, ASPO, October, 1948, pp. 35-42.

[14] John Rannells, *op. cit.*, pp. 189-190.

[15] Hans Blumenfeld, "The Form of the Metropolis." A paper presented before a meeting of the Philadelphia Housing Association, May 20, 1958. (1717 Sansom Street, Philadelphia 3, Pa.)

changes in regard to population and employment, but which also shows specifically informed concern for the lengths of work trips required. Rather than considering the changing extent or "sprawl" of metropolitan areas as an independent variable in the indefinite lengthening of work trips, it is proposed that we think of controlling the length of the work trip by offering economical alternative modes of travel and by research-based efforts to achieve desirable distributions of residential density and efficient workplace locations in the metropolitan structures of the future.

Glossary

Since ideas are preserved and transmitted by words, we cannot perfect language without perfecting science, nor science without perfecting language; however certain may be the facts, however accurate may be the ideas that spring from them, they will nevertheless convey only a false impression unless we have exact terminology with which to express them.

Chemistry
by Antoine Lavoisier, eighteenth-century chemist

It has been found desirable to adapt terminology and to propose a few new terms in order to deal in detail with work travel as measured in O-D and motor-vehicle-use surveys. All trips referred to are those made by individuals (person-trips) rather than those made by the vehicles.

Three major sources were relied upon for terminology not adapted or proposed in this report. The sources were:

1. The Glossary in Mitchell and Rapkin, *Urban Traffic, a Function of Land Use,* pages 213-219. A number of concepts given in that book and defined in its Glossary are not redefined here, though utilized in the text of the present report.

2. U.S. Department of Commerce, Bureau of the Census. *United States Census of Population:* 1950. Volume II, Characteristics of the Population; Part I, U.S. Summary. Also, Volume III, Selected Population and Housing Characteristics. U.S. Government Printing Office.

3. U.S. Department of Commerce, Bureau of Public Roads. *Manual of Procedures for Home Interview Traffic Study.* Revised edition, October, 1954, reprinted 1957 by the Public Administration Service.

One other source was found helpful.

4. U.S. Department of Commerce, Office of Area Development. "Central Business Districts and Their Metropolitan Areas . . . ," November, 1957.

1. Census week: "The 1950 data on employment status pertain to the calendar week preceding the enumerator's visit. This week, defined as the 'census week,' is not the same for all respondents, because not all respondents were enumerated during the same week. (A majority of the population was enumerated during the first half of April.) . . ." 2*
2. Central business district: A district arbitrarily determined for planning purposes as the locus of the greater part of commercial activities within a city and which constitutes an area of very high land values. In it are located many stores, service establishments, offices, hotels, and theaters. It is an area of high daily accumulation of persons and vehicles. (*Adapted from source 4.*)
3. Dwelling unit: "In general, a dwelling unit is a group of rooms or a single room, occupied, or intended for occupancy, as separate living quarters by a family or other group of persons living together or by a person living alone." 2
4. Employed: "Employed persons comprise all civilians 14 years old and over who, during the census week were either (1) 'at work'—those who did any work for pay, or profit, or worked without pay for 15 hours or more on a family farm or in a family business; or (2) 'with a job but not at work'—those who did not work and were not looking for work but had a job or business from which they were temporarily absent because of vacation, illness, industrial dispute, bad weather, or layoff with definite instructions to return to work within 30 days of layoff. Also included as 'with a job' are persons who had new jobs to which they were scheduled to report within 30 days . . ." 2
5. Employee density: A ratio obtained by dividing average (or normal) number of major-shift production (or hourly pay) workers by occupied acreage of land area of an industrial site. Office workers are excluded from the ratio. (Occupied area includes transportation facilities of the site itself, but excludes totally unused or reserve land. Areas specifically set aside for landscaping or other aesthetic reasons are not considered as unused.)
6. Employment density: A ratio obtained by dividing average (or normal) number of major-shift workers of all occupations by occupied acreage or groundspace of a workplace site.
7. Family: "A family, as defined in the 1950 Census, is a group of two or more persons related by blood, marriage, or adoption

* The number following the definition refers to the source in the above list.

and living together; all such persons are regarded as members of one family."[2]

8. Gainful workers: ". . . persons reported as having a gainful occupation, that is, an occupation in which they assisted in the production of marketable goods, regardless of whether they were working or seeking work at the time of the census . . ."[2]
9. Government workers: "Persons who worked for any governmental unit, regardless of the activity which the particular agency carried on."[2]
10. Head of household: "One person in each household is designated as the 'head.' . . . The head is usually the person regarded as the head by the members of the household. Married women are not classified as heads if their husbands are living with them at the time of the census."[2]
11. Home-work trip: A trip made from residence to a workplace by a person going to perform gainful employment.
12. Household: "A household includes all the persons who occupy a dwelling unit. A person living alone in a dwelling unit or a group of unrelated persons sharing the same living accommodations as partners is counted as a household."[2]
13. Journey to work: The two systems of movement in a city or metropolitan area, comprising all "home-work" trips and all "work-home" trips.
14. Labor force: ". . . includes all persons classified as employed or unemployed, . . . and also members of the armed forces. . . . The 'civilian labor force' comprises the employed and unemployed components of the labor force."[2]
15. Median: The value which divides a statistical distribution into two equal parts: one half of the cases being less than this value and one half of the cases being greater than this value.
16. Net residential density: A ratio obtained by dividing the amount of total resident population by the total of the acreage of parcels used wholly or predominantly for residential purposes, including use for dwellings with stores. Streets, vacant lots, and all other uses are excluded.
17. Not in labor force: "Persons not in the labor force comprise all civilians 14 years of age and over who are not classified as employed or unemployed, including persons doing only incidental unpaid family work (less than 15 hours). Included in this group are persons primarily engaged in their own home housework, students, seasonal workers in an 'off' season, the retired, persons unable to work, inmates of institutions, and persons not reporting on their employment status."[2]
18. Origin-destination (O-D) traffic survey: See Appendix C, Section C.11.

19. Other-work trip: A trip made from work or other nonhome location in connection with regular activities of a person's gainful employment.
20. Primary wage-earner: Same as head of household, except in cases where wife is employed and husband is not.
21. Private wage and salary workers: "Persons who worked for a private employer for wages, salary, commission, tips, pay-in-kind, or at piece rates." [2]
22. Resident employee: ". . . employees of the head of the household who usually reside with their employer, and their relatives residing in the same household. . . . In 1950, the small number of relatives of resident employees were included in the count of resident employees with whom they lived . . ." [2]
23. Secondary wage-earner: Employed member(s) of household other than primary wage-earner. (Does not refer to resident employees.)
24. Self-employed workers: "Persons who worked for profit or fees in their own business, profession, or trade, or who operated a farm either as an owner or tenant . . ." [2]
25. System of movement: A broad pattern of movement that is functionally related to an organized activity of business or other social action.[1]
26. Trip: Travel from one place to another to accomplish a particular purpose. Except for some types of recreational trips in which travel is sought for its own sake, the purpose is to be accomplished after termination of the trip. A trip is made by a single mode of transportation. In general, movements of less than two city blocks in length are not regarded as trips. For purposes of metropolitan surveys, intercity travel by public carrier to or from local terminals is excluded from the meaning of the term, although local travel to or from the terminal is included. Transportation of children of four years of age or less is disregarded.
27. Trip-end: A trip origin or a trip termination. These terms refer to the anatomy of the trip itself rather than to the place of occurrence.
28. Trip purpose: See Appendix C.2 for definition of terms designating the purpose of the trip in O-D surveys.
29. Unemployed: "Persons 14 years old and over are classified as unemployed if they were not at work during the census week but were either looking for work or would have been looking for work except that (1) they were temporarily ill, (2) they expected to return to a job from which they had been laid off for an indefinite period, or (3) they believed no work was available in their community, or in their line of work." [2]

30. Unpaid family workers: "Persons who worked without pay on a farm or in a business operated by a member of the household to whom they are related by blood or marriage. The great majority of unpaid family workers are farm laborers." [2]
31. Work-home trip: A trip to a person's residence from a place where his gainful employment is pursued.
32. Work travel: The three systems of movement in a city or metropolitan area, comprising all "home-work," "work-home," and "other-work" trips.
33. Work trip: A work-purpose trip as defined in connection with Bureau of Public Roads format O-D surveys. This is a trip to location of place of gainful employment, or to locations which must be visited in the course of an ordinary day's work.[3]

Appendix A

Locating Districts of Employment in Cities

THE LOCATION AND SIZE OF VARIOUS CENTERS OR DISTRICTS OF EMPLOYMENT are of considerable importance in transportation planning. If gross plans are to be prepared, it may be necessary to specify the size and location of only the major centers. On the other hand, if more careful plans are in process, it is desirable to know the districts of employment throughout an urban area, arranged in analytically useful geographic divisions. This appendix describes three types of studies which are useful for this general purpose.

A.1. LOCATING MANUFACTURING EMPLOYMENT

A study prepared in Philadelphia recently had as its principal objective the determination of the present and probable future location of employment in manufacturing industry. Some attention was given to estimating total employment, but this was subsidiary to the purpose of the report, "Industrial Land and Facilities for Philadelphia." [1]

Primary source of the data used in locating 1948 manufacturing employment was the *Twelfth Industrial Directory for the Commonwealth of Pennsylvania,* published in 1950 by the Bureau of Statistics, Department of Internal Affairs, Commonwealth of Pennsylvania. Other data were obtained directly from employers, particularly in cases where several sites were occupied by a particular employing firm.

Total volumes of employment for manufacturing industries to-

[1] Prepared for the Philadelphia City Planning Commission, at the Institute for Urban Studies, University of Pennsylvania, 1956.

gether with the street addresses of firms are listed in the Directory. The data apply to manufacturing industries of Standard Industrial Codes 20 through 39, as well as to certain miscellaneous activities such as laundering. This miscellaneous activity was dropped from the compilation. Addresses of the plants were coded by census tract and block, using a census tract index and a map from the 1950 Census of Housing. For all large employers with more than one plant, a direct check was made to ascertain the amount of employment at each plant location. In addition, certain omissions were found in the Directory listing, and data on this employment were added. The error in the data supplied is considered to be within 5 per cent.

A.2. LOCATING EMPLOYMENT BY MANUFACTURING AND NONMANUFACTURING INDUSTRY

Another example of a located-employment study is that prepared for the Philadelphia Urban Traffic and Transportation Board in August, 1955, published as *Technical Memorandum Number 3*. Table 21 of the Memorandum showed the 1950 distribution of jobs throughout the Philadelphia metropolitan area by "Planning Analysis Sections." These sections are rather broad geographic units. In Philadelphia, which has 12 sections and about 125 square miles of area, they average over 10 square miles each and they are somewhat larger outside the city. Table 22 gave estimates for manufacturing and nonmanufacturing employment in 1980 in the same area, allowing for effects of improved transportation in the intervening years.

In this study, the total number of jobs in Philadelphia were approximated from studies of numbers of employed residents, as reported in the Census of Population and other data relating to employment. The distribution of jobs within Philadelphia was based on the location of industrial employment as prepared by the Industrial Land Study, on information from the Census of Business for 1948 (which indicates employment location data for a number of businesses and services), on Planning Commission reports (*Philadelphia Central District Study*, 1951, and *Economic Base Study*, 1949) and on other employment data.

Finally, proportions were established for groupings of employment volumes for various industries, based on consideration as to whether they are primarily market-oriented (department stores), other-industry-oriented (transportation), or located according to population density (grocery stores).

Certain of the data sources contained duplicating information. Where this occurred, to the extent possible, duplications were eliminated. The multiple listing of some items provided a nettlesome kind of indication of the completeness of coverage and accuracy of some of the source material.

A.3. LOCATING EMPLOYMENT BY INDUSTRY

Another valuable type of located-employment study is that of a format suggested by the U.S. Bureau of Employment Security. In this, the attempt is made to distribute all employment in a city or metropolitan area within large arbitrary districts, by industry. A distribution is also prepared to indicate the residences of workers in each industry, within the same district boundaries, according to Census data. It is not known for how many cities such studies have been made, but reports of this type are available for the New York metropolitan area (1950), Chicago (1952), and Philadelphia (1946).[2]

A study of this type yields two tables: one which indicates the number and proportion of residents in each of the districts, by industry of occupation, and another which shows the location and number of workers in each district by industry of occupation. From such a study, the relative surplus or deficit, by districts and by industry, can be determined. Thus the minimum import or export of labor by industry becomes determinable, but no complete thread of connection may be found between the sets of data for resident workers and that for located jobs. If the links among the two sets of data can be found, this information would be very helpful in giving scale to journey-to-work estimates. The study prepared for the New York metropolitan area by the State Department of Labor, Division of Employment, Bureau of Research and Statistics, is most thoroughgoing and yields data tabulated for a considerable number of types of industries. In general, the New York study found tremendous imbalance between employment and residence in almost all counties of the metropolitan area. New York City as a whole was found to have about 91 per cent of all metropolitan area employment and about 84 per cent of the residence.[3] It cannot be inferred that the

[2] A highly detailed survey yielding located-employment data was conducted in Philadelphia in 1956 as a part of the National Housing Inventory. See Table II-8 in particular for additional references to the 1956 survey.

[3] In the New York report for 1950 it was also found that all listed sections except Manhattan had more employed persons by residence than by place of work. But Manhattan had 64 per cent of the located jobs and only 23 per cent of the resident workers in the metropolitan area. Here are the brief summary statistics which underlie the several million manhours per day spent in the New York metropolitan journey-to-work. More recent data may be found in one of the studies sponsored by the Regional Plan Association, E. M. Hoover and Raymond Vernon, *Anatomy of a Metropolis:* The Changing Distribution of People and Jobs within the New York Metropolitan Region (Cambridge: Harvard University Press, 1959).

7 per cent difference in these two figures is the amount of commutation into New York City to fill these jobs. It is larger than this because an undetermined number of New York City residents travel out of that city to jobs elsewhere. This indeterminacy requires supplementation by traffic information concerning the type of trip between the places of residence and employment.

Appendix B

Estimating the Interchange of Workers among Cities

APPROXIMATING THE INTERCHANGE OF WORKERS AMONG OUTLYING AND central cities of a metropolitan area is of interest for a number of reasons. An approximation of this type helps to give scale to more localized or finely detailed studies of relationships between residence and workplace. Another facet of the problem is this pertinent financial question: How many persons who pay real estate taxes or rent in suburban areas earn their livelihood and receive police, fire, and other municipal services in the central city? The converse is also interesting. These questions are of great importance to industrial and governmental economists who seek to determine the distribution of personal income payments within a metropolitan area. Such distributions are part of the studies of the tax base which are vital in surveys of the location and financing of various kinds of public facilities.

The example below is presented for illustration of a technique for relating traffic and economic data, which may prove useful to planners and urban economists. It was prepared by the writer for use by the City of San Francisco, with data applying as of July, 1947.[1] The estimate of worker interchange was later found to be reasonably complete when considered in terms of reliability of the original data.[2]

The net number of commuters to San Francisco in July, 1947, as estimated from available vehicular and passenger traffic data, was 70,000 (Table B-1). Commuters were considered to be workers who

[1] This was a product of the Transportation Technical Committee, whose staff was furnished by De Leuw Cather and Company.

[2] See *Daily Trips in San Francisco*, San Francisco Department of City Planning, 1955. The Bay Area Metropolitan Traffic Survey was conducted in 1947, but complete data were not available when this study was prepared.

TABLE B-1

NET ACCUMULATION OF COMMUTERS[a]

SAN FRANCISCO, JULY, 1947

Item	Route and station	Inbound		Outbound		Net accumulation		Source
		Total passenger vehicles	Total passengers	Total passenger vehicles	Total passengers	Passenger vehicles	Passengers	
1	Ferry Building		566		159		407	1
	S.F.—Oakland Bridge							
2	First Street[b]	3,146	4,717	1,230	1,846	1,916	2,871	1
3	Fifth Street[b]	3,499	5,211	2,544	3,761	955	1,450	1
4	Key Transit vehicles		8,541		655		7,886	1
5	Out of town buses		997		906		91	1
6	Key System Railroad		12,289		2,072		10,217	1
7	Southern Pacific		9,293		342		8,951	2
8	Greyhound—7th Street		6,417		1,573		4,844	2
9	Greyhound—Sansome Street		2,826				2,826	2
10	San Mateo Interurban Streetcar (40) at County Line[e]		240		240			3
11	Hillsdale and F, Colma[e]	50	81	178	288	−128	−207	4
12	Junipero Serra Blvd. SM Co.[e]	468	759	822	1,330	−354	−571	4
13	Skyline Blvd.[e]	Negligible Traffic						4
14	Ocean Shore Highway[c]	159	258	81	131	78	127	4
15	El Camino Real[d]	694	1,120	731	1,180	−37	−60	4
16	Golden Gate Bridge[e]	1,757	2,950	758	1,270	999	1,680	4
17	Bay Shore Highway	2,865	4,650	1,340	2,170	1,525	2,480	4
	SUBTOTALS		60,915		17,923		42,992	

18	Students, University of California	1,000
19	Daly City, Colma, Bayshore town	3,400
20	Tourists leaving S.F.	600
21	Commuters starting work 7:15 a.m. or earlier (.80 × 9,350)	7,480
22	Commuters, parttime workers starting work 9:15 a.m. or later (.80 × 4,350)	3,480
23	Night-shift commuters (.80 × 7,065)	5,650
24	Night-shift commuters leaving S.F. through cordon (.80 × 131)	100
25	Commuters with day off	2,260
26	Commuter absentees	2,090
27	Commuters on vacation	1,390
	TOTAL	70,442

[a] Sources:

1. *Cordon Count Data for the Metropolitan Traffic District.* 1947. Transportation Technical Committee of the Department of Public Utilities, Public Works, City Planning, and Police, City and County of San Francisco.

2. *Origin-Destination Survey of Persons Entering and Leaving San Francisco Via Southern Pacific Railroad and Greyhound Bus, January, 1948.* Transportation Technical Committee of the Departments of Public Utilities, Public Works, City Planning, and Police, City and County of San Francisco. Compiled by R. M. Bartle and H. S. Lapin.

3. *Origin-Destination Survey of Passengers on San Mateo Interurban Streetcar Line, March, 1948.* Transportation Technical Committee of the Departments of Public Utilities, Public Works, City Planning, and Police, City and County of San Francisco.

4. *Division of Highways, Department of Public Works*, State of California. Bay Area Metropolitan Traffic Survey. February 19, 1948.

[b] 1.47 to 1.52 Persons per vehicle.

[c] 1.62 persons per vehicle.

[d] 1.68 persons per vehicle.

TABLE B-2

COMMUTERS ENROUTE TO WORK BUT OMITTED FROM TRAFFIC COUNTS
SAN FRANCISCO, JULY, 1947

Industry	*Estimated ratio of commuters to total workers*	*Total number off-shift workers*	*Night shift hours*	*Correction for night workers "leaving" in cordon count*	*Day shift hours*	*Day shift workers starting work 7:15 a.m. or earlier*	*Day shift commuters (from Column 7)*	*Night shift commuters (from Column 3)*	*Correction commuters (from workers in Column 5)*
1	2	3	4	5	6	7	8	9	10
Manufacturing	.20								
Food:									
(*a*) Bakeries		1,000	2p.m.-10p.m.		6a.m.-2p.m.	2,000	400	200	
(*b*) Sugar		150	3p.m.-11p.m.					30	
		100	11p.m.-7a.m.	100	7a.m.-3p.m.	1,200	240	20	20
Printing, publishing		500	Around	250	N.A.	N.A.		100	50
Paper		50	4p.m.-11:30p.m.		8:30a.m.-4p.m.			10	
Chemicals		100	4p.m.-12m.		8a.m.-4p.m.			20	
Furniture		100	4:15-12:45a.m.		7:45a.m.-4:15p.m.			20	
Primary metals		60	3p.m.-11p.m.					12	
		15	11p.m.-7a.m.	10	7a.m.-3p.m.	400	80	3	2
Fabricated metals		210	3:30p.m.-12m.					42	
		90	12m.-8:30a.m.	45	8:30a.m.-3:30p.m.			18	9
Electrical machinery		100	4p.m.-12m.		8a.m.-4p.m.			20	
Machinery (except electrical)		100	4p.m.-12m.		8a.m.-4p.m.			20	
Transportation equipment		1,500	4p.m.-12m.					300	
		500	12m.-8a.m.	250	8a.m.-4p.m.			100	50
Other industries		1,000				15,000	3,000	200	
TOTAL		5,575				18,600	3,720	1,115	131

Transportation, communications, utilities	.20							
Transportation		10,000	1*	1*	9,000	1,800	2,000	
Communications		2,500	1	1	4,000	800	500	
Utilities		800	1	1	800	160	160	
TOTAL		16,300			13,800	2,760	2,660	
Trade	.20							
Wholesale		1,000	2*	2*	4,000	800	200	
Retail								
Food and dairy		1,000	2	2	3,000	600	200	
Eating and drinking places		4,000	1	1	1,000	200	800	
Motor vehicles and filling stations		700	2	2	700	140	140	
Other retail		400	2	2			80	
TOTAL		7,100			8,700	1,740	1,420	
Service	.10							
Business service, auto storage		900	2	2	500	50	90	
Personal service								
Domestic		1,000	1	1	2,000	200	100	
Hotels and lodging		4,000	1	1	2,000	200	400	
Laundry, cleaning		1,300	2	2	1,500	150	130	
Amusement		3,000	1	1			300	
Professional and related service		1,000	2	2	300	30	100	
TOTAL		11,200			6,300	630	1,120	
Government not elsewhere classified	.25	3,000			2,000	500	750	
GRAND TOTAL		43,175			49,400	9,350	7,065	131

*1. Workshifts are staggered around the clock.
*2. Complete information not available.

crossed the San Francisco City line to reach their places of employment. This figure represented the number of persons who came to San Francisco to work, less the number of San Franciscans who worked elsewhere. Considering the methods of gathering data and the possible error inherent in them, it may be said that the net number of commuters may have been greater than this figure, but probably not less.

Tables B-2 and B-3 represent work-sheets for calculations and estimates employed in the study, and Table B-1 presents the summary of calculated and known data. Commuting workers were considered in two principal groups, those employed in the regular daytime working hours and those employed in the late afternoon and night on the off-shifts. Coherency of the assumptions made was partially substantiated in the comparison of several cross-sectional proportions. Figures were prepared for the net accumulations of workers in San Francisco taken as proportions of total employed in the City, by each shift and for the total employed force.

RECOMMENDATIONS

The summation of *net* commuting workers, the first such attempt in the San Francisco Bay Area, was based on available statistics in combination with quantitative estimates. The former are more reliable, of course, and were used wherever possible. With this in mind, it is recommended that the following be made a part of the data regularly summarized by public agencies, in order to facilitate reliable estimates of net commuting workers.

1. Proportion of absentee employees, by industry.
2. Proportion of employed persons beginning work at 7:14 or earlier in the morning; proportion beginning work from 7:15 to 9:14 (or some other end-point datum 10 to 15 minutes following the modal time-of-arrival period); and proportion beginning work at 9:15 or later through the midday period; proportions employed on swing and graveyard shifts. Compilation by industrial categories noted in tables.
3. Number of transportation workers and others regularly on duty outside the central city, whose employment is listed at a workplace in the central city.
4. Proportion of employed persons with days off on weekdays, by industry.

TABLE B-3

Commuters with a Scheduled Day Off, Monday Through Friday San Francisco, July, 1947

Industry	*Estimated ratio commuters to total workers*	*Workers with day off during week*	*Daily average workers off during week*	*Daily average commuters off during week Column 2 and Column 4*
1	2	3	4	5
A. Transportation, Communications, Utilities	.20			
Transportation		19,000	2,720	540
Communications		5,000	720	140
Utilities		500	70	10
B. Trade	.20			
Wholesale		8,000	1,330	270
Retail		30,000	5,000	1,000
Personal service		20,000	3,330	670
Amusement		5,000	830	170
C. Government, not elsewhere classified	.25	4,000	570	140
TOTALS		91,500	14,570	2,940
Total of persons not included in traffic counts, Table 1, or other calculations			11,200	2,260

Note: Groups A and C, Column 4, represent 1/7 of Column 3.
Group B, Column 4, represents 1/6 of Column 3.

The people who help to maintain around-the-clock services and utilities constitute a considerable portion of the off-shift workers listed in Table B-2. The number of these persons was divided by 7 to obtain a daily average. In trade the employment was divided by 6 to obtain the daily average.

Therefore, the totals of the night workers (Table B-2, Column 3) and day workers (Table B-2, Column 7) beginning work by 7:15 a.m. were compared to total employment in those industries.

For industries in Table B-3:

Off-shift workers	32,800
Day workers (at work by 7:15 a.m.)	30,000
Total employment in these industries	273,800

$$\frac{30{,}000 + 32{,}800}{273{,}800} = 23\%$$

Thus the off-shift and early morning workers in Table B-2 account for 23 per cent of the employment in the industries listed in Table B-3, and it is logical to Conclude that 23 per cent of the workers in Table B-3 are accounted for in Table 2. The totals of Columns 4 and 5 were reduced by 23 per cent to eliminate the duplication.

Appendix C

Work-Trip Data: Sources, Terminology, Precision, and Completeness

SOURCES FROM WHICH INFORMATION ON THE JOURNEY TO WORK MAY BE obtained, typical terminology used in these sources, and the precision and completeness of the data are described in this Appendix.

C.1. SOURCES OF JOURNEY-TO-WORK INFORMATION

The primary source of information is in the origin-destination surveys as conducted by state highway departments according to procedures established in large measure by the U.S. Bureau of Public Roads. Other sources include surveys of employment by state departments, motor-vehicle-use surveys conducted by the Bureau of Public Roads, studies by city or regional planning commissions, travel investigations by marketing researchers, data provided by individual companies, and traffic counts conducted in cities. Additional data which supplement these may be obtained from census publications, summaries of local employment from such sources as chambers of commerce and departments of employment, from local transit companies and from industry associations such as the American Transit Association and the Automobile Manufacturers Association.

The systems of work travel in cities provide a framework in which all these data from a variety of sources may be related. Information from all sources described above are utilized in the present study. To enhance understanding of these materials, attention is given in this chapter to discussions of these source materials.

Origin-destination surveys are among the most complete social surveys of any kind made in urban areas. As such, they represent virtual treasuries of information about the city. It is believed that the full value of such expensive surveys in most instances has not been gained by local or other agencies.[1]

The usual procedure has been to compile data on total trips between zones, and sometimes total trips by mode, for use in planning a few major routes or possibly an integrated complex of streets and highways. The full information, however, as contained in the detailed zone-to-zone information by purpose of trip, by mode, by land use of zone of origin, has often been left on the original punch-cards. Another loss occurs in the failure to provide enough subtotals in the final reports to permit extraction of data for subsequent use.

It is hoped that the present study may have some effect in changing this needless waste by indicating purposes and ways for further utilization of information from these sources. A major consideration is that by additional tabulations, particularly those run on an area-of-origin basis, knowledge may be gained which may usefully be related to localized demographic and economic information.

C.11. Origin-Destination Traffic Surveys

A great amount of valuable information concerning the physical patterns of work travel can be obtained from origin-destination (O-D) surveys. These studies determine the structure of travel throughout an urbanized area by means of a large sample of home interviews and traffic counts at selected stations. A cordon line is drawn on a map of the area to be studied. In the field, motorists may be interviewed at all points where highways cross the cordon line. Within the cordon, the internal survey is conducted by first dividing the urban area into units, in diminishing size order known as sections, districts, and zones. A systematic sample[2] of the population, varying upward from 3⅓ per cent in very large cities to as much as 20 per cent in cities of about 50,000 population, is determined on the basis of tolerable error and the considerations of probability sampling, and dwelling units throughout the city are

[1] Exceptions to the above statement would include recent work of the Bureau of Public Roads and the California Division of Highways, and the large-scale survey efforts completed or still under way in Detroit, Chicago, Pittsburgh, and Philadelphia.

[2] A systematic sample is one in which every nth unit, such as every 30th or every 5th unit, is included for interviewing. The selection is made to equalize to the extent feasible the probability of including each family unit or each dwelling unit in the survey.

NOTE: It is believed that data-sifting such as is suggested herein, in combination with sophisticated approaches to area probability sampling will lead to less costly data-gathering procedures in future travel surveys.

selected for interviews. The data obtained are collated and expanded to the level indicated by proportions of the sample taken.

Three possible methods may be used to check the correspondence between expanded sample results and actual traffic volumes. These are (1) screen-line comparison, (2) control-point comparison, and (3) cordon-line comparison. Usually either the screen-line or the control-point comparison is selected as the major control procedure, and the cordon-line comparison only supplements whichever of the other two methods is used.

A screen line is a physical barrier of considerable length in the city, such as a river or railroad right-of-way. Trips in the survey which show origins and destinations on opposite sides of the line are compared in volume with actual counts of the traffic crossing it.

A control point is a key focus in a city, such as a major bridge or underpass. If this method is selected, a question is included in the interviews about any trips made past the control point. Data from the expanded sample which show trips past the point are compared with actual ground counts taken at the location.

In the cordon-line comparison, data from the internal survey concerning trips to and from the external area are compared with total volumes recorded in the external survey.

After any necessary adjustments are made in the expanded sample data, they are compiled from the punch-cards and tabulated in ways of interest to transportation engineers, city planners, and others concerned.

C.12. Motor-Vehicle-Use Studies

Sampling surveys of travel characteristics are being conducted in a number of cities and states according to the format designed by the U.S. Bureau of Public Roads. These surveys do not provide locational or origin-destination information, but rather considerable detail on the kinds of trips made, their volumes and lengths, and the people who make the trips. The surveys are somewhat misleadingly titled "motor-vehicle-use studies," although all modes of urban travel, including walking, are considered. As such, the sampling surveys constitute descriptive studies of the total "market" of urban travel, minus the detail of internal trip distribution or interchanges.[3]

Published summary data for urban travel in six states have been studied for the present effort. Data for states other than the original six have become available more recently. The information sum-

[3] Similarly, travel characteristics of Londoners are measured from time to time to study work trips and other urban travel. The February, 1954 sampling was taken for the Road Research Laboratory and the London Transport Executive. Other similar studies in London have been made in 1938-39 and in 1948.

marizes travel characteristics and information on the use of vehicles for cities in Arkansas, Louisiana, North Dakota, Oklahoma, South Dakota, and Wisconsin.[4] The data are presented by population-size class of cities.

One of the tabular presentations indicates the influence of the one-way length of the trip on the selection of mode of travel. Composite figures for cities of over 100,000 population in the six states showed fairly equal usage of public transportation and autos for work trips of from one to five miles length. In the five-to-ten-mile range, however, auto trips were double the number of public transportation trips, and in the ten-to-twenty-mile range, they were triple the number of those by public carriers.

Differences in the mode of travel for work trips selected by each occupational group are also indicated. For instance, it was found in the same cities that 31 per cent of the skilled workers rode transit facilities in getting to work, while 47 per cent of unskilled workers and 65 per cent of personal-service employees used public transportation for such trips.

The hope for improving urban transportation lies in obtaining greater understanding of the structure of urban travel. Relating trip characteristics to the travelers, as well as to the focal points of the journeys themselves, is an important key to that understanding.

C.13. Studies of Place of Residence and Place of Work

A valuable but skeletal outline of the home-work relationship may be obtained by comparing data from two independent sources: the Census areal distribution of the labor force, by industry-of-occupation, and the list of located employment, by industry, as prepared from sources discussed in Appendix A. A procedure for this type of comparison has been prepared by the U.S. Employment Service. Studies based on the procedure have been carried out in several major metropolitan areas. Three of these were examined in connection with the present report: those for the New York, Chicago, and Philadelphia labor markets. The New York report is considered to be outstanding.

In at least one state, a similar type of study has been attempted in order to assess the relative surpluses and deficits in the labor force among the component counties. The State of Indiana Employment Security Division analyzed information concerning the residences of employees of 1,067 firms.[5] The data indicate the proportions of the

[4] T. A. Bostick, R. T. Messer, and C. A. Steele: "Motor-Vehicle-Use Studies in Six States," *Public Roads,* XXVIII (December 1954).

[5] *Survey of Commuting Patterns, March-April 1950,* July 1950, 141 S. Meridian Street, Indianapolis.

workers who reside in the county of their respective workplaces and the proportions who work in other counties.

The information cannot very well be subjected to distance analysis because of the varying locations of the workplaces within the counties and because the counties, of course, are not comparable in area. (This is the same problem experienced in the London analysis, as described by Westergaard. See Chapter I.) Though the state information applies only to employment covered by unemployment insurance and primarily that in manufacturing industry, it still might be enlightening to pursue comparisons between available traffic studies, such as that prepared for Fort Wayne[6] in Allen County, and the results of the state worker-residence survey.

C.2. TERMS USED IN O-D SURVEYS

The following is a listing of the classifications of trip purposes as used in O-D surveys according to the format established by the Bureau of Public Roads.[7] To make coding manageable, only one purpose, the primary motivation, is considered for each trip.

Definitions of other terms of relevance to the present subject are given in the Glossary.

Purpose	*Description*
Work	Trip to location of place of employment, or to locations which must be visited in the course of an ordinary day's work. "Trips made by a doctor in making his calls and by a salesman calling on prospective customers are classed as trips to work."
Business	Trip "to complete transactions not considered part of a person's regular employment." A trip to a bank, for instance, which is not part of a person's regular job duties, is classed as "business."
Medical-dental	Trip for consultation about health with professionals. Does not refer to trips made by professionals to see patients; such trips would be considered to be work purpose.

[6] State Highway Commission of Indiana. *Traffic Survey Report, Metropolitan Area Traffic Survey. Fort Wayne, Indiana.* 1945. Also, Traffic Audit Bureau report on Fort Wayne, and Allen County, Indiana, 1945.

[7] The information below is adapted from this publication of the U.S. Bureau of Public Roads: *Manual of Procedures for Home Interview Traffic Study*. October 1954. See pages 52-54 and 80.

School	Trip by a registered student to an educational institution.
Social-recreation	Trip for cultural purposes or for a visit to friends or sources of organized recreation.
Change travel mode	Trip to location where change is made to another mode of transportation enroute to the ultimate destination. There is a tendency in some recent surveys to drop this category, since it reflects a mechanical characteristic of the trip, rather than its purpose.
Eat meal	Trip to obtain a regular meal (not for obtaining refreshments, a trip for which purpose would be treated as recreational). A trip to home for a meal before the day's work is completed is an "eat meal" rather than a home trip.
Shopping	Trip to inquire about or to purchase goods or personal services. Trips to radio repair shops or barber shops, e.g., are classed as shopping, as are trips to retail stores.
Serve passenger	A trip made in an automobile to transport a rider from one specific location to another.
Home	Trip to the residence address where the home interview is conducted. This class is used in internal surveys only. In the external survey, a person interviewed at a station enroute to home from work would be listed as on a work trip.

Of particular interest in this report is the definition of work trips. These comprise three systems of movement as defined in the Glossary. From the point of view of studying the relation between homes and workplaces, the home-work system of trips are generally simpler to study and of greater use for analytic purposes than the trips from work to home. Distinction must be made between trips from work to home in the internal and external surveys. Whereas trips made from work to home in the internal survey are classed as home trips, in interviews for the external survey the homeward trips are given the purpose of the activity from which they originate. Thus, in the external surveys, the trips to home from workplaces are labeled as work trips.

The term "journey to work" has been given a specific meaning in this report. A definite term was needed to relate O-D quantitative information on systems of work-movement to the great diurnal mass movements between homes and workplaces. Work travel is the most general term, since it includes all trips made in connection with a person's gainful employment.

Another important distinction concerning work trips was quantified in the Detroit study (1953), that between the first and later trips from home or other origin to work. As a growing number of workers take second jobs or make multiple home-work trips in a given day for other reasons, then greater variations will be made more likely in trip generation ratios compiled on a per capita or a per resident worker basis. Current data indicate that Philadelphia's total home-work trips are expanded 5 per cent by persons holding more than one job.[8] Thus, to increase the predictability of the data, it is important in studying the generation of trips from residential areas that tabulations be made of first home-work trips, as compared with other work trips from residence.

Following is a typical listing of the modes of travel by which it is possible to describe trips in O-D surveys. Not all these categories may be applicable in particular situations. In most cases, it is to be expected one or more of modes numbers 3, 5, 8, or 9 in the "Suggested List" will be found inapplicable. This is important because procedures in routine machine analysis make 10 modes preferable to 11.

Present Practice	*Suggested List*
1. Auto driver	1. Auto driver
2. Auto passenger	2. Auto passenger
3. Streetcar	3. Streetcar
4. Bus	4. Bus, local
5. Streetcar-bus	5. Bus, express
6. Taxi passenger	6. Taxi passenger
7. Truck passenger	7. Truck passenger
8. Train	8. Train
9. Subway-elevated	9. Rail rapid
10. Ferry	10. Walk
	11. Other

In strict terms, some of the above categories are not so much modes of travel as they are classes of travel, but for purposes of standardization all the above are considered the modes of urban travel.

The distinction of local bus travel from express bus travel will grow in importance, it is believed, as the urban expressway systems are developed in metropolitan areas. If one of the listed categories is found not applicable, thus creating a blank in the list, it would

[8] In the Current Population Survey in Philadelphia, interviews indicated that 7 per cent of employed men and 2.5 per cent of employed women held two jobs during the week ended July 13, 1957. Together, these multiple job-holders comprised 5 per cent of all employed persons during the survey week.

be desirable to separate trolley coaches from other buses, to aid in transit route planning. The fourth mode in the list of present practice, streetcar-bus, has sometimes been interpreted as referring to trips consisting of consecutive rides on the two types of vehicles. This interpretation is inconsistent with the definition of a trip as a movement made by a single mode. Part of the confusion stems from the fact that the same vehicle is described by three terms: streetcar-bus, trolley-bus, and trackless trolley.

C.3. ASPECTS OF MEASUREMENT

Precision of the results of a sampling study refers to the closeness of matching between what can be reproduced from the sample and the information that would be found if a total census were made, using all the same methods of measuring, interviewing, etc., as in the sampling study. Measurable precision is important in assessing the reliability of the study. It also allows judgment of efficiency; thus, an efficient sample design is one which yields relatively high precision per unit of expenditure.

The difference between the description based on the sample results and the actual value is the *accuracy* of the sample study. We are primarily concerned with accuracy and try to measure it where possible, but it is precision which is likely, at best, to be all that is measurable in some types of sampling studies.

Since accuracy is a term involving an absolute concept, the *actual* value of a quantitative entity, it is likely not to be measurable in sampling surveys, for the practical consideration of cost. It may, therefore, be more meaningful to use the term *completeness* in referring to the comparison between expanded sampling information and field data which give an indication of the size of the population being measured.

Another concept we need to consider is *relevance* of the data being measured to the objectives and needs of the study. We may measure work trips with high precision and high accuracy, but if the definition of work trips is too broad, then our findings may still not be useable. Such is not the case, if care is used in interpretation. However, it is true that the term *work trips* includes three systems of trips: (1) all home-work trips, (2) all other-work trips, and (3) external survey work-home trips. Unless we refine our data to distinguish among these three systems, the results may lack relevance for the given objectives. Indeed, one of the conclusions of the present study is that the importance of the subject for city and transportation planning calls for clear distinction among the three systems of trips associated with regular gainful employment.

C.31. Checking Completeness of Work-Trip Data

Work trips are considered to be the most accurately reported of the trips for all purposes considered in urban origin-destination studies. This point is illustrated in the usual procedure in dealing with O-D data; where adjustments are found necessary to make the expanded sampling information accord more closely with counts obtained at the screen-line or control points, the smallest proportional adjustment is made on volumes of work-purpose trips. This may be roughly inferred from a graph showing the hour-by-hour comparison of expanded sample data for trips crossing a screen line with the actual screen-line counts (Figure C-1). In the peak periods—thus, at the times of work trip dominance—much the closest agreement between sampling and control information is found. Further, the morning peak, being more exclusively composed of work trips than the evening peak, is usually the more accurately reported of the two periods.

In the adjustment of expanded-sample data described above, full agreement between sampling and control data may be too difficult a goal to seek. To illustrate, experience in the Omaha survey of 1956 was reported this way:

> Peak-hour travel, which includes the work trips, was found to be very well reported. Off-peak travel, not so completely reported, was subject to special adjustments to bring trip interview data within 90 per cent of ground counts at the screen lines.[9]

Thus, a total matching was not sought. Unless the screen-line counts represent average volumes of those taken over the entire period of interviewing—which may be several months—it is probably just as well not to regard the screen-line data as an absolute standard. Overadjustment of survey data is possible, to the point where confidence may be lost in applicability of the results.

Screen-line and control-point checks are included in O-D surveys to provide a measure of the discrepancy between the expanded-sample information and actual field measurements of trips for all purposes. However, these comparisons only lead to inferences concerning correspondence of the expanded-sample information and the total actual travel distribution. The screen-line and control-point checks are themselves a kind of sampling procedure relative to total urban travel. Thus, accuracy of the O-D information is not measure-

[9] Howard, Needles, Tammen, and Bergendoff; and Wilbur Smith and Associates, *Omaha Metropolitan Area Proposed Trafficway System*, Vol. I, October 1956, p. 47.

Figure C-I

COMPARISON OF SCREEN-LINE GROUND-COUNT DATA WITH EXPANDED SAMPLE DATA (AUTOS & TRUCKS) PHILADELPHIA - CAMDEN - 1947

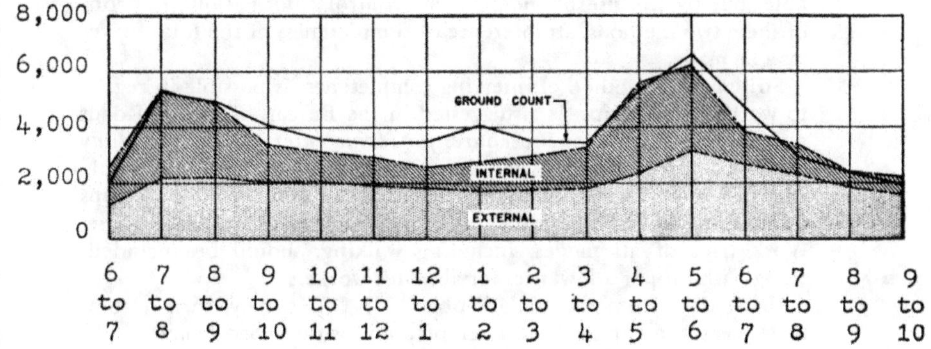

Delaware River Screen Line

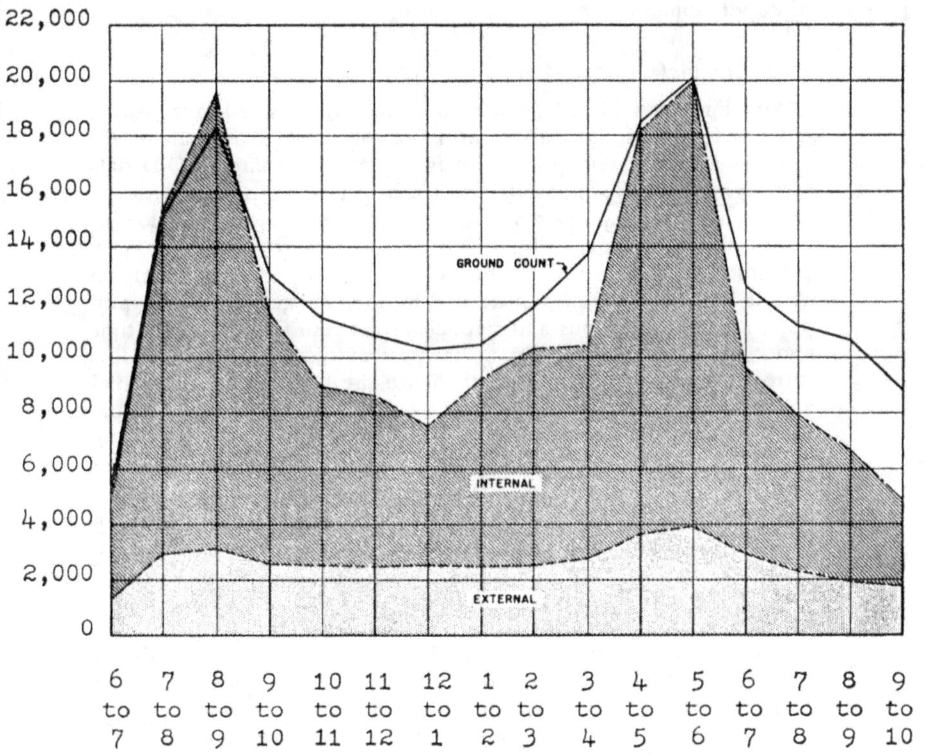

Schuylkill River Screen Line

Source: Pennsylvania Dep't. of Highways, et al. Philadelphia - Camden Area Traffic Survey - 1950

able, but by the use of independent control information from one of these two methods, an inference of completeness of the total survey may be made.

Still another method of inferring completeness is possible in regard to work-purpose trips. It is suggested in the Bureau of Public Roads publication, *Manual of Procedures for Home Interview Traffic Study* (p. 87). This method calls for the selection of a zone in which only workplaces are located and comparison of the number of work trips terminating in that zone with the number of persons employed there. Work trips of all modes, including walking, should be included, along with proper allowances for absentee workers.

Using data from the Philadelphia 1947 O-D survey, such a complete measurement has been attempted and is described below. Additionally, however, work trips originating from internal residential areas were compared with the total numbers of residents gainfully employed away from home, as indicated by data from the Census of Population.

C.32. Completeness of Work-Trip Data from Origins

A major source of independent information for small areas within cities is found in the tract information of the U.S. Bureau of the Census. To the extent possible, it is desirable to compare O-D data for origin work trip-ends with the census information. Difficulties in making such a comparison are encountered because of the following factors which introduce disparities: persons employed at residences; persons who walk to work; multiple work trips per person; persons absent from work; confusion caused by the change-travel-mode purpose; and miscellaneous and implicit errors resulting from the time and methods of data collection. Each attempted adjustment of data introduces some error of its own; hence the total error of the final answer may be too great to leave any real meaning in the actual numbers derived. Nevertheless, an illustration of the process of comparison is believed to be desirable in order to aid in the design of future surveys for greater ease in comparison studies based upon independently collected data. The following example pertains to seven Philadelphia residential areas for which O-D data were gathered in 1947 and Census data in 1950.

C.321. Persons Employed at Residences. As shown in Table C-3, it was assumed that one fourth of the self-employed and all the unpaid family workers worked at home in 1947. This assumption results in an estimate that 2.6 per cent of all Philadelphia employed work at home, and about 2.8 per cent for the areas in the present example. A 1956 residential sampling study in Philadelphia indicated that 3.9 per cent of the heads of households worked at home. Relative to the total gainfully employed force, the proportion for heads of households is likely to be high.

C.322. Persons Who Walk to Work. A figure of 8 per cent was taken as the proportion of the local employed force who walk to work on an average day. Data for Philadelphia on this proportion are shown in Table C-1.

TABLE C-1

Year of survey	*Reporting group*	*Walkers as percentage of group*
1934	Main wage-earner	22.8
1952-53	All gainfully employed	7.6
1956	Employed head of household	9.3

Apparently the number of walk-to-work trips has been decreasing over time. (The difference between the proportions of 7.6 and 9.3 per cent is not considered significant.)

C.323. Multiple Work Trips per Person. Multiple work trips may be made during the course of a business day, by salesmen, for instance, or by other persons who may hold more than one regular job. A third source of multiple work trips may be persons who leave their work-places for other than official reasons, and whose return trips to work are reported as work trips.

To gain a sense of the proportion of other-work trips, the trips in the Philadelphia internal area (1947) classed as "from work-purpose to work-purpose" were studied relative to total internal work trips. The mode of the work trip was found to be related to the proportions of such trips. Considering the several aspects of the problem as described below, the following estimates were employed in the present study: 80 per cent of the total work trips made by drivers and passengers in personal vehicles (autos, trucks, taxis), and 98 per cent of those made by persons traveling to work by mass vehicular modes were considered to be first trips to each job.

The following figures are shown to indicate detail in the estimated proportions. Let a figure A equal total vehicular internal work trips, and B equal A minus the number of vehicular internal work trips "from work-purpose to work-purpose." Then B/A represents the proportion of first vehicular work trips to total vehicular work trips. The Philadelphia-Camden internal area B/A ratio in 1947 was calculated at 0.84. In Detroit (1953) it was calculated at 0.83. (See page 124 of the DMATS *Data Summary and Interpretation* for the original data.) Another item from Detroit, a memo titled "Work Travel in the Detroit Area," indicated that the weighted average first vehicular work-trip ratio for that city was 0.765. The difference between the 0.83 and 0.765 ratios probably represents persons who held more than one job and persons who reported multiple trips to a single workplace from nonwork origins during a normal workday. Further details from the Detroit memorandum are shown in Table C-2. One additional fact was noted in it (not shown in Table C-2),

TABLE C-2

RATIO OF FIRST TRIPS TO ALL VEHICULAR WORK TRIPS

INTERNAL SURVEY AREA, DETROIT, 1953[a]

Mode of travel	*Total trips to work*	*First trips only*	*Percentage first trips*
Auto driver	820,944	547,108	66.5
Vehicle passenger	189,081	170,543	90.0
Transit passenger	255,835	250,325	97.9
TOTAL	1,265,860	967,976	76.5

[a] From a memo titled "Work Travel in the Detroit Area," by J. D. Carroll, Jr. (mimeo.) 2 pp. About 1955. (Preliminary data.)

that the number of first walking trips to work in Detroit was 63,325, or 6.1 per cent of all first trips to work in the internal area.

As noted previously in the Appendix (see footnote 8), 5 per cent of employed persons in Philadelphia (1957) were enumerated as having two paid jobs. In arithmetic attempts to reconcile located job data with figures on work destinations, the multiple-job factor will be implicit and no correction need be made. However, in attempting to match located residences of workers with the origin of work trips (Table C-3), an adjustment will have to be made for the factor of multiple employment. Consideration of the level of economic activity over time will be necessary in using this factor in any one city or area. (In 1947, in Philadelphia, the multiple-employment factor was probably somewhat less than 5 per cent.)

There is no reason to believe that a complete separation exists among the several factors: e.g., it would seem likely that some of the people with multiple jobs travel directly from one to another, and hence are represented in the group of "work-purpose to work-purpose" trips. The lack of specificity on this matter in the usual

O-D data adds to the value of the information on first trips gathered in Detroit.

C.324. Persons Absent from Work. Total absentees on a typical workday are estimated at about 6 per cent of the employed force. For Philadelphia in 1952-1953, it was reported that 9.1 per cent of the gainfully employed did not travel out-of-doors (motor-vehicle-use survey, preliminary information). In studies for Linn County, Iowa, and Allen County, Indiana, it has been reported that only about 90 per cent of employed persons went out-of-doors on a typical weekday. If 3 per cent of these were persons working at home, then a figure of 6 to 7 per cent represented those who were on vacation, sick, or otherwise absent from work.

C.325. Other Factors. The O-D information was collected three years prior to the enumeration of the 1950 Census of Population. This is not considered to hold much potential error in the present case, since the seven areas of the example were selected for their relative stability in total population, as well as for distribution throughout the city.

Trips reported "from home to change-travel-mode" could obscure as much as 12 per cent of the home-to-work trips, as data for the total Philadelphia-Camden internal area indicate. However, the proportion is not likely to be quite this high in any one area, since some of the mode changes both of local residents and of work-travelers passing through the particular area would be made in each traffic area-unit, and would thus add to the number of locally originating work trips. But the number of work trips beginning in the more purely residential areas, particularly in those served by bus transit only, is more likely to be underreported because of the work trips broken by a change of mode en route.

Of the seven residential areas, data for three show noticeable inconsistencies in Table C-3. These are a large residential area in South Philadelphia (B-3); a small area in the northeast part of the city (K-5); and an old area in the northwest, in the Germantown area (I-4). Data for the South Philadelphia area appear considerably underreported, unless the unlikely assumption is made that major changes occurred in employment between 1947 and 1950, or that considerably more than 8 per cent of the employed walked to work. Perhaps both possibilities existed to some degree; it is also quite possible that underreporting occurred in the interview process itself, not an uncommon phenomenon in lower-income residential areas.[10] The overreporting for K-5 may be due partly to the small size of

[10] Discussion of the possibilities of bias from this source is given in a paper presented before the American Statistical Association, September 13, 1957, by Robert H. Hanson and Eli S. Marks, "The Influence of the Interviewer on the Accuracy of Survey Results."

TABLE C-3

Comparison of Numbers of Resident Employed Persons with Origin Work Trip-Ends Philadelphia, Average Weekday, 1947

Subsection	Miles and direction from center city	Net residential population density (per acre)	Gainfully employed[a]			Calculated				O-D
			Self	Family unpaid	Total	Total[a,b] travelers to work	First[a,c] vehicular work trips	First[d] trip factor	Total vehicular work trips	Reported[e] vehicular work trips
B-3	2.3, S	203	5,912	209	59,126	57,689	49,890	0.948	52,500	42,870
D-3	3.4, W	126	2,160	29	19,149	18,580	16,068	0.892	18,000	18,334
D-4	4.3, W	129	2,905	33	19,127	18,368	15,885	0.912	17,400	19,067
F-1 + F-5	4.3, N	156	1,955	46	31,060	30,525	26,398	0.921	28,600	26,878
I-4	7.8, NW	32	2,013	16	13,195	12,676	10,962	0.884	12,400	15,845
K-2	7.8, NE	83	3,907	33	26,507	25,497	22,050	0.870	25,400	25,396
K-5	10.0, NE	36	228	5	2,216	2,154	1,863	0.841	2,200	3,398
TOTALS									156,500	151,788

[a] Based upon 1950 Census of Population.

[b] Assumes that 25 per cent of self-employed and all family unpaid workers work at home.

[c] Assumes that 6 per cent of travelers to workplaces are absentee or on vacation on an average day, and that of those who leave home to work, 8 per cent walk.

[d] Assumes that first work trips per employed person per job held are represented by the following proportions by mode: drivers and passengers in private vehicles, 80 per cent, passengers in mass modes, 98 per cent.

[e] From O-D survey of 1947. In the period 1947-1950, there was not much growth in the areas selected, with the possible exception of K-5. Had there been considerable growth in K-5, it would have been in the direction of reducing the discrepancy between the 1947 and 1950 figures.

that part of the sample and partly to trips listed as "change-travel-mode to work" from origins in that area. The apparent error in area I-4 (sixth highest in median income of the 67 subsections of the city), overreported work trips relative to resident workers, is likely to be due to the latter cause, since a number of persons drive to and park at stations of the commuter rail lines serving northwest Philadelphia.

It is to be seen that the totals of calculated and actual work trips are within 3 per cent. Not too much can be made of this close agreement other than to notice that some compensating errors appear to be involved, and the general point that there is greater reliability in dealing with relatively large volumes of the data represented in an expanded-sample survey.

C.33. Destination Work Trip-Ends in the CBD

Trips to workplaces in the Philadelphia CBD appear to account for an employment of about 300,000 in 1947. The only available figures for CBD employment in the post-war period are considerably higher than 300,000, but there is no assurance that the latter estimate is any more reliable than the work-trip data. This is known: at least one control study (screen line) is available for the O-D figures to check their completeness. Shown in Table C-4 are the available data on located employment.

TABLE C-4

ESTIMATES OF TOTAL EMPLOYMENT IN PHILADELPHIA CBD

Year	*Employment*
About 1928	293,188[a]
1934	223,772[a]
1950	387,000[b]

[a] Philadelphia *Real Property Survey*, 1934. Figure of 293,188 is listed as the maximum at an unspecified year prior to 1934. The unspecified year is guessed to be 1928.

[b] Philadelphia Urban Traffic and Transportation Board, Technical Memorandum No. 3, page 144. Prepared by Marketers Research Service, Inc., 1956.

In assembling trip-end information, it was assumed that one tenth of internal work trips to CBD destinations were multiple work trips (or other-work trips), the volume of which does not relate directly to numbers of jobs. Work trips to the CBD from external areas by auto were assumed to be in the same proportion to total external trips as was found for external auto work trips to all destinations in the internal area (43.6 per cent). This figure differed from an estimate in U.T.T.B. Memorandum No. 4, noted in Table C-5. Additional

TABLE C-5

PERSON WORK TRIP-ENDS IN PHILADELPHIA CBD, BY ORIGINS, 1947[a]

Trip origin	Person-trips: Work purpose	Person-trips: Total
Internal area, walkers	35,000 (est.)	n.a.
Within CBD	24,000	76,000
Internal area (outside CBD)	212,700	470,000
External area, auto	10,200 (est.)	23,300
External area, rail and bus	25,000 (est.)	32,000
TOTAL	306,900	

[a] Sources: Internal area: Philadelphia City Planning Commission, *Person-trips to the Central City*. 1954; External area: Philadelphia Urban Traffic and Transportation Board, Technical Memorandum No. 4; Estimates of the present study are indicated. N.a. connotes "not available."

assumptions were that walkers constituted 12 per cent of the total employed, and that absentees numbered 6 per cent of the total employed. The assembled trip-end information is shown in Table C-5.

The general indications of the data available would seem to be that the work trip-end information, particularly as reported in the O-D survey for vehicular trips, is more reliable than data on located employment obtained from other sources.

From the screen-line comparison of trips over the Delaware River (bounding the east side of the CBD), we would expect the peak-period trips to be well within 5 per cent of the ground counts or of total counts at key nodal points. The O-D survey tells us "for the 16-hour period studied, the expanded interview data reproduced 93.3 per cent of the auto and truck trips crossing the Cooper River."[11] The Cooper River is located on the Camden side of the Delaware

[11] Pennsylvania Department of Highways, *et. al., Philadelphia-Camden Area Traffic Survey*. Vol. 1, p. 4.

River, and is tributary to the latter. Thus, the underreporting of trips for all purposes over these two screen-lines was measured as 6.7 and 13.6 per cent, respectively. The hourly diagram, Figure C-1, shows such good peak-period correspondence of data that reported work trips, at least over the Delaware, must have been within the margin of 5 per cent of the actual daily average at the time of measurement. A similar control on representativeness of data by purpose of trip, other than what is suggested in methodology in this chapter, does not exist.[12] At any rate, with a discrepancy of the order of magnitude of that observed here, it appears that some further study is in order. Though the employment estimate for the CBD of close to 400,000 is likely to be too high, it also appears that the employment estimate of 300,000 based upon trip-ends is too low. Whatever the details may be in the case of the Philadelphia CBD, the general method of relating the two independent types of information is believed to be useful to those who measure such urban phenomena.

C.34. Check of Work Trips Classified by Industry

In the O-D interviews, work trips are classified by industry-of-occupation of the employed persons. If this information were relevant as gathered and reasonably complete, it would be helpful in gaining understanding of the geographic distribution of the labor force. In order to test relevance of the data classified by industry-of-occupation, a test was made of the count of trip-ends for "manufacturing" industry to seven selected employment centers in Philadelphia. These were selected on the bases of (1) being outside the CBD, (2) being in representative industrial locations, (3) the traffic district boundaries could be matched with planning analysis sections for which manufacturing employment in 1948 was available. As shown in Table C-6, the data on internal work trip-ends were compared with the figures on located employment. (Located employment figures, as listed in Source *a* of the table, conform with a particular definition of "manufacturing industry," industrial groups 20 through 39 in the Standard Industrial Classification of the U.S. Bureau of the Budget.) The job data for 1948 were close enough in date to compare with the O-D data, gathered in the latter part of 1947, to discount the possibility of major discrepancies because of changes over time.

It may be seen that the results of the test were far from satisfactory.

[12] A chance of low probability exists that the 50,000-plus trips "from change-travel-mode to work" to CBD destinations were not included in the deck of work-trip cards run. This would also be a relevant point for further inquiry in the Philadelphia-Camden origin-destination survey of 1947.

TABLE C-6

COMPARISON OF O-D DATA ON WORK TRIPS TO MANUFACTURING ESTABLISHMENTS WITH LOCATED MANUFACTURING EMPLOYMENT

PHILADELPHIA, 1947 AND 1948

Subsection	*Traffic district*	*Employment manu-facturing industry 1948*	*Internal work trip-ends, manu-facturing 1947*	*Total internal work trip-ends 1947*
B-7	083	13,815	7,054	19,495
D-9	051	272	398	7,587
G-1	012	8,628	3,766	9,288
G-3	014	13,851	7,258	11,508
J-2	036	4,595	2,155	13,313
K-1	015	8,821	6,287	20,385
L-4	028	6,200	1,153	2,587
TOTAL		56,182	28,071	84,163
ADJUSTED[a]			34,600	

Sources: *Industrial Land and Facilities for Philadelphia*, Institute for Urban Studies, 1956. p. 147; Special tabulation of 1947 O-D data, prepared for present study.

[a] Allowing for external trips, walking, absentees, and 5 per cent work-to-work trips. See footnote 13.

The sum of the unadjusted trips to internal work destinations for the seven districts was only half of the volume indicated from independent data on located manufacturing employment. Even after adjustment, the work trip-end figures stood at about two thirds of the volume to be expected from the located employment data for SIC manufacturing categories.[13]

Consideration must be given to why, in the present comparison,

[13] Total work trips to manufacturing destinations were adjusted from 28,071 to 34,600 on consideration of a negative allowance for multiple work trips not related to total employment (5 per cent), of 8 per cent walkers of those residing in the internal area, of external trips composing 12 per cent of total direct vehicular trips, and of 6 per cent of total employed being absent on an average weekday.

the data on work trips may be less complete than would be expected from the apparently nearly complete reporting of peak-period trips (mostly work trips) as indicated in the hour-by-hour graphed data of the screen-line tests. Just a segment of total work trips was under consideration, and that segment was derived from answers to a fairly abstract question on industrial classification made by members of the general public, comprised in large part of housewives. Accurate transmission of this information depends heavily upon the interpretive abilities of the general public.

On the other hand, the located manufacturing employment data were derived from professional sources. They were assembled from private and governmental records and cross-checked for duplication, in many cases directly with the employers. The question remains, of course, as to whether the SIC manufacturing category (codes 20 through 39) is completely comparable with the O-D code for manufacturing industry. The latter question is considered below.

It would be unwise to draw a conclusion about the value of the industry-of-occupation information of O-D surveys from only one study. Since a question of representativeness of the data obtained has been raised, it will be necessary to consider the matter more fully. This will be done by comparing Bureau of the Census information on industry-of-occupation of residents of urbanized areas corresponding roughly to the internal areas of three origin-destination surveys: Madison (1949), Minneapolis-St. Paul (1949), and Philadelphia-Camden (1947). In this way, not only may any error peculiar to the Philadelphia-Camden survey be isolated, but also the problem of comparing data based upon the SIC manufacturing code with O-D manufacturing code data is avoided. (Of course new problems of industrial classification are introduced, but there is somewhat greater control in this second method, since the 27 industrial categories of Table 35 in Volume II, *Characteristics of the Population,* of the 1950 Census can be fairly well fitted to the industrial codes of the O-D format. Only a few minor problems of comparability were believed to apply: these are noted in footnote *a* of Table C-7.)

Study of the distributions for Philadelphia-Camden, Madison, and Minneapolis-St. Paul internal surveys brings out the fact that underreporting of manufacturing work trips is by no means general. In the latter two surveys, manufacturing work trips appear to be well reported.

On the other hand, several disparities were common to all three surveys. Work trips to the construction industry were overreported in all three. Since the Bureau of the Census gives no breakdown of its construction industry category, it would seem desirable that study be made of the instructions for code 2. The difference between Census and O-D proportions probably lies in the "related maintenance" categories used in the O-D classification.

TABLE C-7

Percentage Distributions of Employed Residents by Industry Groups

Philadelphia-Camden, 1947 and 1950

Code	Industry classification[a]	Census data 1950 Residents of Philadelphia: Urban[b] place	Census data 1950 Residents of Philadelphia: Urbanized[c] area	O-D data 1947[d] Residents of Phila-delphia-Camden internal area	Urban area per cent minus O-D per cent
2	Construction	5.60	5.90	8.16	−2.26
3	Manufacturing	35.21	35.63	26.67	8.96
4	Transportation and Communication	8.34	8.38	8.83	−0.45
5	Wholesale and retail trade	26.84	26.53	29.24	−2.71
6	Personal service	8.36	7.82	5.03	2.79
7	Amusement and recreation	0.82	0.84	0.71	0.13
8	Professional and related	6.51	6.60	11.56	−4.96
9	Government	6.32	6.26	6.57	−0.31
X	Other	0.28	0.42	2.96	−2.54
Y	Not reported	1.72	1.62	0.27	1.35

[a] Industries are grouped according to industry codes shown in the *Manual of Procedures for Home Interview Traffic Study*. U.S. Bureau of Public Roads. Manual 2B, Public Administration Service. Edition of 1954. pp. 73-74. Codes 0 and 1 are grouped with code X. Census classifications are distributed according to the O-D codes. Educational services-private are listed in code 8 and educational services-public are listed in code 9. There was no way of removing Postal Service employees and employees of government-owned utilities from code 9 and placing them in code 4. Not much error appears to result from the code 9-code 4 displacement between Census and O-D classifications.

[b] Corresponds to Philadelphia City; employed persons totaled 827,243.

[c] Roughly corresponds with internal area in 1947 survey, but includes a somewhat greater urban area than the internal area. Employed persons totaled 1,154,867. Data of *b* and *c* from *Census of Population: 1950*, Vol. II, Characteristics of the Population. Part 38, Pennsylvania. U.S. Bureau of the Census, 1952, Table 35.

[d] From special tabulations of work-purpose trips of this report, based upon results of O-D survey by Pennsylvania Dept. of Highways, New Jersey State Highway Department, and local agencies in cooperation with the U.S. Bureau of Public Roads.

Industry classifications for which O-D data were overreported in all three surveys also included:

Code 4 Transportation, communication, and other public utilities; and

Code 5 Wholesale and retail trade including finance, insurance, and real estate.

Classifications for which O-D data were underreported in all three surveys included:

Code 6 Personal service; and

Code 7 Amusement, recreation, and related services.

Discrepancies in the Madison survey of over 3 per cent of the total occurred in codes 2, 5, 6, 8, and 9. The discrepancy in code 9, government employment, was most notable. The problem here might have revolved around the heavy employment at the University of Wisconsin, the State University, which is assigned by the Census to the public education category. In the O-D classification, it is probable that a good part of this employment was listed under code 8, professional and related services, and University maintenance personnel were likely to have been listed under code 2, construction and maintenance.

Where disparities were observed in all three surveys between Census and O-D trip data by industry, it would seem worthwhile to restudy the industrial classifications of the O-D procedure before designing a survey. The categories in which this occurs are O-D industrial codes 2, 4, 5, 6, and 7.

Other disparities noted in all three surveys appear to have been peculiarities of the individual efforts. The reasons for such disparities (for example, that noted for the Philadelphia Manufacturing work trips) appear to derive from the interview process. The "industry-of-occupation" question is considerably more difficult to answer than such a question as "Where did this trip end?"

In conclusion, the comparisons of Tables C-7 and C-8 would seem to indicate that industry-of-occupation can be gathered with a fair degree of reliability in the O-D process (consider manufacturing work trips in Madison and in Minneapolis-St. Paul), but that the apparent reliability has varied by type of industry in individual surveys. Considerably greater attention to errors introduced by misunderstanding of industrial classifications when designing the survey and conducting the interviews appears to be warranted.

TABLE C-8

Percentage Distributions of Employed Residents by Industry Groups Madison and Minneapolis-St. Paul, 1949 and 1950

Code	Industry classification[a]	Census data 1950 Urbanized areas		O-D data 1949 Residents of internal areas	
		Madison[b]	Minneapolis-St. Paul[c]	Madison[d]	Minneapolis-St. Paul[e]
2	Construction	5.84	5.41	9.80	6.93
3	Manufacturing	16.73	25.54	16.05	26.52
4	Transportation and communication	6.58	11.14	7.33	10.99
5	Wholesale and retail trade	28.15	32.26	32.55	34.48
6	Personal service	8.47	6.68	3.28	3.94
7	Amusement and recreation	1.17	1.10	0.55	0.89
8	Professional and related	10.05	8.38	13.89	6.19
9	Government	21.15	7.76	11.16	6.51
X	Other	1.19	0.60	5.39	2.88
Y	Not reported	0.67	1.13	—	0.67

[a] See note *a*, Table 7 preceding.

[b] *Census of Population: 1950*, Vol. II, Characteristics of the Population. U.S. Bureau of the Census. Part 49, Wisconsin, Table 35. The urbanized area corresponds roughly with the survey area. 1950 population of the urbanized area was 110,111, and 1949 population of the survey area was estimated at 107,000. Employed persons totaled 48,135 in 1950.

[c] *Loc. cit.*, Part 23, Minnesota, Table 35. The urbanized area corresponds roughly with the survey area. 1950 population of the urbanized area was 985,101, and 1950 estimated population of the internal survey area was 952,925. Employed persons totaled 415,349 in 1950.

[d] From a special tabulation prepared for the present report, based upon the O-D survey conducted by State Highway Commission of Wisconsin, City of Madison in cooperation with U.S. Bureau of Public Roads. Work trips tabulated totaled 38,346. An additional 4,281 work trips of the internal survey had insufficient detail to be tabulated; these were not listed under Code Y above in order not to warp unduly the other percentages, since the other two cities considered had much smaller proportions of Code Y trips.

[e] From a special tabulation prepared for the present report, based upon the O-D survey conducted by Minnesota Department of Highways, Cities of St. Paul and Minneapolis in cooperation with U.S. Bureau of Public Roads. Work trips tabulated totaled 386,059.

C.35. *Work Trip Accumulations throughout the City*

It may be considered desirable to apportion the destinations of work trips by major areas in order to obtain an independent check upon available estimates of located employment throughout a city. Further, it may be considered desirable to attempt a matrix analysis of relationships among originating work trips and job opportunities at destinations. In any case, it is believed that the full value of this type of analysis of the systems of work travel has yet to be realized and further work on work-trip accumulations is recommended.

Trip-card analyses run by area of destination, by mode, and by time of arrival, for internal and external trips to the areas of interest, are suggested. A destination analysis for Philadelphia, based upon the internal survey, is shown in Table C-9. The data on time and mode can be of aid in determining requirements for public transportation and parking areas, when small areas are considered.

The importance of work-trip data by place of employment may be seen in the fact that at least two metropolitan areas, Los Angeles and Toronto, have utilized work-trip surveys taken *at the places of work* as guides in making major transportation decisions. A third such survey is described below. The primary reason, in the two cases named, was the economy of measuring only the journey to work and of having employers obtain work-trip data from their employees. The information was obtained from cards distributed to the employees. The precision of these surveys does not appear to be measurable, and the accuracy of residential data was probably not susceptible to determination. Only the proportion of response was calculable, and while a high proportion of response lends credence to the data, it does not guarantee accuracy. But, per dollar invested, both the Los Angeles and Toronto surveys were worthwhile as pilot studies in situations where probability sampling at residences was infeasible or too expensive for the funds available.

The third employee-postcard study mentioned was conducted as a follow-up to gain understanding of new traffic patterns created by a major new highway facility.[14] This example concerned the traffic effects of Route 128, a circumferential bypass highway around the Boston metropolitan area. By gaining information at workplaces located along the new facility, it was possible to select a population with a high rate of usage of Route 128, as well as some control figures (gate-counts and employment volumes) to indicate completeness of the survey.

These three surveys lead to the question of whether or not it would be desirable to conduct randomly distributed interviews at

[14] A. J. Bone and Martin Wohl, "Economic Impact of Route 128," *Traffic Engineering*, XXVIII (July 1958), 11-17, 36-39.

TABLE C-9

Work Trips to Workplaces and Estimates of Located Employment (in thousands)

Philadelphia, 1947, 1948, and 1950

Section	Trips to workplace,[a] internal survey	Adjusted work trip-ends[b]	Employment estimates: Manufacturing[c]	Non-manufacturing[d]	Total[e]
A (CBD)	236.7	301.3	59.5	336.3	395.8
B	77.7	85.9	32.9	31.8	64.7
C	19.6	21.7	12.0	11.9	23.9
D	82.9	91.7	13.2	45.0	58.2
E	113.8	125.8	68.5	49.5	118.0
F	82.8	91.5	63.2	28.5	91.7
G	53.0	58.6	50.5	17.5	68.0
H	7.2	8.0	4.9	4.0	8.9
I	31.9	35.3	7.6	22.8	30.4
J	33.5	37.0	11.7	21.7	33.4
K	54.7	60.5	24.6	62.9	87.5
L	6.2	6.9	7.7	1.9	9.6
City	800.0	924.2	356.3	633.8	990.1

[a] From 1947 O-D survey. Summarized from tabulation of work trips prepared for the present study.

[b] Sections B through L. Assumptions: 85 per cent of internal work trip-ends are directly related to the volume of employment; 8 per cent of workers from the internal area walk to work; internal vehicular trips represent 88 per cent of the total including external trips; 6 per cent of the total are absent on an average workday. Section A, the CBD: See C.33, "Destination Work Trip-ends in the CBD."

[c] Manufacturing employment of SIC codes 20 through 39. From Appendix Table 30, *Industrial Land and Facilities for Philadelphia*, Institute for Urban Studies. 1956. Data for year 1948.

[d] Workers in nonfarm, nonmilitary, nonresidential establishments, in non-manufacturing industries. From Table 21, *Technical Memorandum No. 3*, Philadelphia Urban Traffic and Transportation Board, 1956. Data for year 1950.

[e] Figures are a composite of information for the years 1948 and 1950. See notes *c* and *d* for sources. Estimated employment for Section A (CBD) does not agree with figure of Table C-4 (387,000) because 1948 rather than 1950 manufacturing employment was used in the table above. No estimates are obtainable for nonmanufacturing employment by section for 1948. Manufacturing data for 1948 do not exclude military employment in this category as do the manufacturing data for 1950.

workplaces, as well as at residences, in the O-D internal survey. Such a procedure might well provide another type of control like that presently provided through the use of screen lines and control points, and in this case, one which would apply to the most important system of the O-D trip information. This measure would also facilitate comparison of trip and employment data in cities where walking trips to work have not been recorded.

Another justification for collection of work-trip data from the persons concerned at their workplaces proceeds from the difficulties of matching O-D work-trip accumulation figures with located employment figures (as in the attempt of Table C-9). Considerable adjustment of the data is necessary to provide comparability, and even the facts necessary for the adjustment multipliers are rarely available.

To recapitulate, a special and possibly limited pilot study of work-trip data is recommended, to be based upon information received from wage-earners at their workplaces. If, for reasons of budget, even such a pilot study is infeasible, then at the very least, it would be desirable to conduct sufficient field study and research to obtain adjustment multipliers so the independently gathered control data (such as Census figures) may be effectively utilized. This effort becomes the more important as attempts are made to keep current the expensive O-D survey information. The independent data could help to maintain reasonable volume-scale in the updated trip estimates. Specific adjustment multipliers are named in footnote *b* of Table C-9.

Considering the importance of the journey to work in planning for urban circulation facilities, the relatively low investment needed to carry out the suggestions above should not be allowed to be a deterrent to the fact-gathering and study necessary to accomplish the objective.

C.4. SUMMARY

This appendix has considered the sources of data from which we measure and judge the journey to work. It has presented current terminology and pointed out some possible refinements in the language of the surveys to define more closely the parts of the journey to work. Finally, it has considered the value of the data in terms of such concepts as precision, completeness, and relevance. Examples were given to show attempts to compare O-D derived information with independent control data. Comparisons indicated close agreement in some cases, and wide deviations in others. Explanations for differences in data were provided so far as possible, and in some instances suggestions were made of a means to increase the completeness and reliability of work-trip data.

Selected Bibliography

1. JOURNEY-TO-WORK STUDIES

ADAMS, LEONARD P., and MACKESEY, THOMAS W. *Commuting Patterns of Industrial Workers.* (Research Publication No. 1.) Ithaca, N. Y.: Cornell University Housing Research Center, 1955. 135 pp.

BEYER, GLENN H. *Housing and Journey to Work.* (Bulletin 877.) Cornell University Agricultural Experiment Station, August, 1951. 19 pp.

CARROLL, J. DOUGLAS, JR. *Home-Work Relationships of Industrial Workers.* Unpublished doctoral thesis, Harvard University, 1950.

———. "The Relation of Homes to Work Places and the Spatial Pattern of Cities." *Social Forces,* XXX (March, 1952), pp. 271-282.

———. "Some Aspects of Home Work Relationships of Industrial Workers." *Land Economics,* XXV (November, 1949), pp. 419-422.

CONKLIN, HOWARD E. *A Statistical Analysis of Employment and Farming Data for Open-country Residents in Chemung and Monroe Counties, New York.* Doctoral dissertation, Cornell University, 1948.

GERARD, ROY. "Commuting and the Labor Market Area." *Journal of Regional Science* I (Summer, 1948), pp. 124-130.

GLASS, RUTH. "London on the Move," *Times* (London), June 18, 19, 1956, p. 11 (both issues).

ILLINOIS STATE EMPLOYMENT SERVICE. *Analysis of Employment by Place of Work and Place of Residence,* April, 1949. 7 pp. 5 tables, 5 charts.

INDIANA EMPLOYMENT SECURITY DIVISION, Research and Statistics Section. *Survey of Indiana Commuting Patterns, March and April, 1950.* July, 1950.

LAWTON, R. "The Daily Journey to Work in England and Wales." (An Analysis of the Report on Usual Residence and Workplace, Census of England, Wales, 1951.) *The Town Planning Review,* XXIX (January, 1959), pp. 241-257.

LIEPMANN, KATE K. *The Journey to Work: Its Significance for Industrial and Community Life.* New York: Oxford University Press, 1944. 199 pp.

NEW YORK STATE DEPARTMENT OF LABOR, Bureau of Research and Statistics. *Residence and Employment in the New York Metropolitan Area. 1950*. August, 1953.

POPULATION ASSOCIATION OF AMERICA, Sub-committee on Journey to Work. *Final Report*. Chapel Hill, N.C., 1957.

REGIONAL PLAN ASSOCIATION, INC. *New York's Commuters*, Bulletin No. 77. July, 1951. 205 East 42nd Street, New York 17, N. Y.

RANYAK, JOHN A. *A Theoretical Approach to the Journey to Work*. Unpublished baccalaureate thesis, Massachusetts Institute of Technology, January, 1952.

SOUTH SIDE PLANNING BOARD. *Employee Transportation Survey. A Study of the Journey to Work*. Spring, 1954. 2018 Calumet Ave., Chicago 16.

THOMPSON, JEAN. "The Journey to Work. Some Social Implications." *Town and Country Planning* (November, 1950), pp. 441-446.

WATSON, JOHN E. "Travelling Time to Work, Some Notes from the New Zealand Census of 1945." *Social Forces* (March, 1952), pp. 283-292.

WESTERGAARD, JOHN. "Journeys to Work in the London Region." *The Town Planning Review*, XXVIII (April, 1957).

WHITING, ROBERT F. "Home-to-Work Relationships of Workers Living in Public Housing Projects in Chicago." *Land Economics*, XXVIII (August, 1952), pp. 283-290.

2. MARKET STUDIES

ALDERSON AND SESSIONS. *Philadelphia Central District Study*. August, 1951.

———. *The Families of the Delaware Valley*. Philadelphia, Pa., December, 1956.

ALEVIZOS, JOHN P., and BECKWITH, ALLEN E. *Downtown and Suburban Shopping Habits Study of Greater Boston*. Boston: College of Business Administration, Boston University, 1954.

BARATZ, MORTON S., and GRIGSBY, WILLIAM G. *Housing Market Analysis and the Journey to Work*. Philadelphia: Institute for Urban Studies, University of Pennsylvania, 1958. Mimeographed.

BRANCH, MELVILLE C. *Urban Planning and Public Opinion*. Princeton, N. J.: Bureau of Urban Research, Princeton University, 1941.

GILMAN, W. C., AND COMPANY. See title under "Traffic Reports."

POLITZ RESEARCH SERVICE, INC. *The Automobile in the Daily Life of the American Population*. For the Automobile Manufacturers Association, April, 1951. 96 pp.

STOP AND SHOP, INC., Marketing Research Department. *Shopping Habits of Super-market Customers*. Boston, Mass., June, 1947. 14 pp.

TRAFFIC AUDIT BUREAU, INC., *Coverage, Repetition and Impact Provided by Poster Showings.* In Cedar Rapids and Linn County, Iowa, 1950. 60 East 42nd Street, New York 17. 26 pp.

———. *Methods for the Evaluation of Outdoor Advertising.* Spring, 1946. 60 East 42nd Street, New York 17, N. Y. 39 pp.

3. POPULATION DISTRIBUTION FUNCTION AND FORM OF CITIES

BLUMENFELD, HANS. "Theory of City Form, Past and Present." *Journal of the Society of Architectural Historians,* VIII (July, December, 1949), pp. 7-16.

———. "On the Growth of Metropolitan Areas." *Social Forces,* XXVIII (October, 1949), pp. 59-64.

———. "The Form of the Metropolis." A paper presented before the Philadelphia Housing Association, June, 1958.

BOURNVILLE VILLAGE TRUST. *When We Build Again.* London: George Allen and Univ., Ltd., 1941. (Concerns City of Birmingham, England.)

BREESE, GERALD W. *The Daytime Population of the Central Business District of Chicago.* Chicago: University of Chicago Press, 1949.

DEWEY, RICHARD. *Peripheral Expansion of Population in Metropolitan Milwaukee.* Unpublished doctoral thesis, University of Wisconsin, 1946.

GREAT BRITAIN, Royal Commission on Distribution of the Industrial Population. *Report.* 1940. M. Barlow, Chairman.

HARRIS, CHAUNCEY D. "A Functional Classification of Cities in the United States." *Geographical Review* (American Geographical Society), January, 1943.

IKLÉ, FRED CHARLES. *The Impact of War upon the Spacing of Urban Population.* Unpublished doctoral dissertation, University of Chicago, 1950.

INDIANA ECONOMIC COUNCIL. "Are the Implications of Factory-Commuting Considered in Your Long-Range Planning?" *News Bulletin,* VIII (November, 1952), pp. 1, 3-4.

INSTITUTE FOR RESEARCH IN SOCIAL SCIENCE, University of North Carolina, Chapel Hill. *Population Distribution-Spatial and Temporal,* September, 1952. 85 pp.

KNEEDLER, GRACE. "An Economic Classification of Cities." *Municipal Yearbook,* 1945, pp. 30-38.

MENGLER, F. A. A. "An Estimate of the Daytime Population of London." *Town Planning Institute Journal,* XXXVIII (March, 1952), pp. 116-120.

STEVENS, GEORGE. "Sample Study of Residential Distribution of Industrial Workers in an Urban Community." *Land Economics,* XXVIII (August, 1952), pp. 278-283.

4. TRAFFIC ANALYSES

BEVIS, HOWARD W. "An Analysis of the Factors Affecting Zonal Interchange Volumes." *Detroit Metropolitan Area Traffic Study*. 1955. 17 pp. Mimeographed.

BLUMENFELD, HANS. *Memorandum re Characteristics of Work Trips*. Metropolitan Toronto Planning Board, 1956.

BONE, A. J., and WOHL, MARTIN. "Economic Impact of Route 128." *Traffic Engineering*, XXVIII (July, 1958), pp. 11-17, 36-39.

BOSTICK, T. A., MESSER, R. T., and STEELE, C. A. "Motor-Vehicle-Use Studies in Six States." *Public Roads*, XXVIII (December, 1954).

CARROTHERS, GERALD A. P. "An Historical Review of the Gravity and Potential Concepts of Human Interaction." *Journal of the American Institute of Planners*, XXII (Spring, 1956), pp. 94-102. This article lists 83 publications on its subject. Hence an extensive coverage of the subject is omitted in the present list.

CHERNIACK, NATHAN. "Methods of Estimating Vehicular Volume with the Aid of the Traffic Patterns." *Proceedings*, Highway Research Board, 1936.

COUGHLIN, ROBERT. *A Projection of Vehicle Trips within Philadelphia*. Philadelphia City Planning Commission, 1957. Mimeographed.

DAVIDSON, ROBERT G. *The Travel of Persons to Downtown Boston*. Unpublished master's thesis, Massachusetts Institute of Technology, 1954.

DODD, STUART C. "The Interactance Hypothesis: A Gravity Model Fitting Physical Masses and Human Groups." *American Sociological Review*, XV (April, 1950).

EVANS, HENRY K. (Ed). *Traffic Engineering Handbook*. Institute of Traffic Engineers, 1950.

FOLEY, DONALD L. "The Daily Movement of Population into Central Business Districts." *American Sociological Review*. XVII (October, 1952), pp. 538-543.

———. "The Use of Local Facilities in a Metropolis." *American Journal of Sociology*, LVI (November, 1950), pp. 238-246.

FOLEY, DONALD L. and BREESE, GERALD. "The Standardization of Data Showing Daily Population Movement into Central Business Districts." *Land Economics*, XXVII (November, 1951), pp. 348-353.

GODLUND, SVEN. *The Function and Growth of Bus Traffic within the Sphere of Urban Influence*. (Bulletin 18, Department of Geography, Royal University of Lund, Sweden.) Lund: C. W. K. Gleerup, 1956.

HIGHWAY RESEARCH BOARD. *Travel Characteristics in Urban Areas*.

(Bulletin 203, January, 1958.) See also HRB Bulletin 224, January, 1959, article by H. S. Lapin.

IKLÉ, FRED CHARLES. "Sociological Relationships of Traffic to Population and Distance." *Traffic Quarterly,* VIII (April, 1954).

LAPIN, HOWARD S. "The Analysis of Work-trip Data . . ." *Traffic Quarterly,* XI (April, 1957), pp. 278-292.

———. "Costs of Automobile Ownership and Use in the San Francisco Bay Area: Sept. 1, 1953." University of California, I.T.T.E., June, 1954. Technical Memorandum B-25. 8 pp. Mimeographed.

———. "Report on Analysis of Urban Work Trips." Highway Research Board, Washington, D.C., January, 1959. (Also in HRB Bulletin 224.)

MATSON, THEODORE M. *War Worker Transportation.* A report to Institute of Traffic Engineers. Yale Bureau for Street Traffic Research, 1943.

MERTZ, W. L., and HAMNER, LAMELLE B. "A Study of Factors Related to Urban Travel." *Public Roads* (April, 1957), pp. 170-174.

MITCHELL, ROBERT B., and RAPKIN, CHESTER. *Urban Traffic, A Function of Land Use.* New York: Columbia University Press, 1954.

OSOFSKY, SAM. "The Multiple Regression Method of Forecasting Traffic Volumes." *Traffic Quarterly,* XIII (July, 1959), pp. 423-445.

OWEN, WILFRED. *The Metropolitan Transportation Problem.* Washington, D.C.: The Brookings Institution, 1956.

RAMÍREZ CARRIL, RAMIRO. "Traffic Forecast Based on Anticipated Land Use and Current Travel Habits." *Proceedings.* Washington, D.C.: Highway Research Board, 1952, pp. 386-410.

SCHMIDT, R. E. and CAMPBELL, M. E. *Highway Traffic Estimation.* Saugatuck, Connecticut: Eno Foundation for Highway Traffic Control, 1956.

SCHNORE, LEO F. *The Separation of Home and Work in Flint, Michigan.* Ann Arbor: Institute for Human Adjustment, University of Michigan, 1954.

WILBUR SMITH AND ASSOCIATES. *Transportation Requirements—Travel Patterns.* Technical Memorandum No. 2, Philadelphia Urban Traffic and Transportation Board, 1955.

STEGMAIER, JOSEPH T. *A Study of the Traffic Generated and the Parking Demand Created by Buildings with Various Types of Space Usage.* Unpublished doctoral thesis, Johns Hopkins University, May, 1948.

VOORHEES, ALAN M. "Forecasting Peak Hours of Travel." A paper presented at Highway Research Board, January, 1958.

———. "A General Theory of Traffic Movement." Institute of Traffic Engineers, 1955.

——— (Ed.). "Land Use and Traffic Models." *Journal of the American Institute of Planners,* XXV (May, 1959), entire issue.

WYNN, F. HOUSTON. "Studies of Trip Generation in the Nation's Capital, 1956-58." Highway Research Board. Washington, D.C., January, 1959.

———. *Traffic Engineering Characteristics*. Vol. 1A. Berkeley: University of California, Institute of Transportation and Traffic Engineering, 1955.

5. TRAFFIC REPORTS

BARKLEY, ROBERT EMMANUEL. *Origin-Destination Surveys and Traffic Volume Studies*. Highway Research Board. Bibliography No. 11. Washington, D.C., 1951.

CALIFORNIA DEPARTMENT OF PUBLIC WORKS, Division of Highways, *et al. Traffic Survey of the Sacramento Area*. 1947-1948. 52 pp.

DETROIT METROPOLITAN AREA TRAFFIC STUDY. Part I. *Data Summary and Interpretation*. July, 1955. J. Douglas Carroll, Jr., Director.

GILMAN, W. C., AND COMPANY. 55 Liberty Street, New York 5, N. Y. *Cleveland Transit System. Special Studies as to Transit Traffic Diverted to or Induced by C.T.S. Rapid Transit*. October 31, 1958.

LONDON TRANSPORT EXECUTIVE. *London Travel Survey. 1949*. London, 1950.

MINNESOTA DEPARTMENT OF HIGHWAYS, *et al. Saint Paul-Minneapolis Traffic Survey*. 1949.

MISSOURI STATE HIGHWAY DEPARTMENT, *et al. A Traffic Survey of Greater Kansas City*. 1944-1945.

OKLAHOMA STATE HIGHWAY COMMISSION, *et al. Tulsa Metropolitan Traffic Survey*. 1944.

PENNSYLVANIA DEPARTMENT OF HIGHWAYS, *et al. Erie Metropolitan Area Traffic Survey*. 1950.

———, in cooperation with U.S. Bureau of Public Roads. *Pennsylvania Motor Vehicle Use Study*. 1952-1953.

———. *Philadelphia Area Traffic Survey*. 1955.

——— AND NEW JERSEY STATE HIGHWAY DEPARTMENT. *Philadelphia-Camden Area Traffic Survey*. 1950.

PHILADELPHIA CITY PLANNING COMMISSION. *Person-trips to the Central City*. 1954.

PHILADELPHIA URBAN TRAFFIC AND TRANSPORTATION BOARD. *Plan and Program 1955*. 1956.

SAN FRANCISCO DEPARTMENT OF CITY PLANNING. *Daily Trips in San Francisco*. April, 1955. 106 pp.

U.S. BUREAU OF PUBLIC ROADS. *Manual of Procedures for Home Interview Traffic Study*. Oct., 1954. Reprinted for distribution by Public Administration Service, 1957.

WISCONSIN STATE HIGHWAY COMMISSION. *Report of the Madison Traffic Survey*. 1951.

Index

(References to Tables are shown in boldface)

www.ingramcontent.com/pod-product-compliance
Lightning Source LLC
LaVergne TN
LVHW011818200926
848484LV00005B/1239